DEDICATION

This book is dedicated to the deeds and exploits, to the sacrifices and achievements of the generation of Australians who proudly served in the RAAF's Phantom, Hornet and Skyhawk Squadrons.

PHANTOM, HORNET AND SKYHAWK

IN AUSTRALIAN SERVICE

Stewart Wilson

ACKNOWLEDGEMENTS

Phantom, Hornet and Skyhawk In Australian Service represents something of a milestone in this series of books as it is the tenth to be published, surely an extraordinary achievement in Australian aviation publishing. And there's a few more to come yet.

In the acknowledgements section of the previous title, I noted: "After what has been a long running and successful series, listing those who assisted may seem to be just a little glib, but rest assured, without them there would not be any books, and their help is appreciated just as much now as it was for the first book of the series".

Those comments will continue to be valid as long as the series is produced, as is the recognition which must be afforded to the dedicated team at Aerospace Publications: publishers Jim Thorn and Margaret Thorn, production manager Maria Davey plus Lee-Anne Simm, Gayla Wilson, Norman Lee and Gerard Frawley.

For this particular book I wish to particularly thank Ross Gillett (RAN PR); Mike Kerr; John Bennett; David Wilson, Bob Piper and Stephen Locke from RAAF Historical; Bill Lines; Mal Lancaster; Roy Phillips; Steve Long; Liz Holtham (RAAF PR); 'George II'; Dave Grant; the Victorian Military Museum; John Allen and Harry Margulies for their artwork; and those individuals whose photographs are acknowledged throughout the book.

STEWART WILSON
Sydney, 1993

Published by Aerospace Publications Pty Ltd (ACN: 001 570 458), PO Box 3105, Weston Creek, ACT 2611, publishers of monthly Australian Aviation magazine.
Production Manager: Maria Davey

ISBN 1 875671 03 X

Proudly printed in Australia by Pirie Printers Pty Ltd, 140 Gladstone Street, Fyshwick 2609
Distributed by Network Distribution Co, 54 Park Street, Sydney 2000
Film by Imagecolour, 178 Sturt Street, Adelaide 5000
Distribution in North America by Motorbooks International, 729 Prospect Ave, Osceola, Wisconsin 54020, USA. Fax: 715 294 4448

CONTENTS

FRONT COVER: McDonnell Douglas F-4E Phantom II 69-0304, 1972; Douglas TA-4G Skyhawk N-13 154647/878 of VC 724, 1979; McDonnnell Douglas F/A-18 Hornet A21-57 of No 75 Sqn in special markings, celebrating the 75th anniversary of the squadron, 1992. (Artwork Z Margulies)

(Drawings not to scale)

ROYAL AUSTRALIAN AIR FORCE

PER ARDUA AD ASTRA

INTRODUCTION

The tenth in our *Australian Air Power* series examines three jet aircraft from the same manufacturer – McDonnell Douglas – although it's worth noting that each aircraft's origins lie with individual manufacturers, McDonnell and Douglas separately (before amalgamation) and in the case of the Hornet, Northrop.

Regarded by many as the "greatest fighter of the modern era", the Phantom is an extraordinary aircraft which was developed for the US Navy as a shipboard fighter but was so good it was also purchased by the US Air Force in very large numbers. Vietnam was the Phantom's war, its versatility coming to the fore and its success complete.

The Phantom was a relatively minor type in RAAF service, 24 examples being leased as an interim measure to fill the gap created by delays in the delivery of General Dynamics F-111s. Phantoms served with the RAAF for less than three years from 1970 but gained many followers during that time. Even today, some question the wisdom of returning the aircraft to the USAF after the lease period expired.

The Phantom section of the book contains by far the most comprehensive photographic coverage of RAAF Phantom operations presented so far.

Developed from the Northrop YF-17 contender for the US Air Force's 'lightweight fighter' competition of the 1970s (won by the General Dynamics F-16), the Hornet was subsequently turned into a naval fighter with McDonnell Douglas taking over prime responsibility for the programme.

Although the Hornet was developed to meet a US Navy requirement, subsequent customers have all operated their aircraft from dry land, including Australia, which selected the aircraft to replace the Mirage III in RAAF service.

A new fighter for the RAAF is always an event of significance, particularly in the modern era as combat aircraft are expected to remain in service for many years. In the book, the background to the Hornet is examined, as is its evaluation and selection by the RAAF, the all important local industry involvement in the manufacturing programme and of course its service.

Completing the trio of McDonnell Douglas products covered by this book is the A-4 Skyhawk, 'Heinemann's Hotrod', the 'Bantam Bomber' and the last fixed wing combat aircraft operated from an aircraft carrier by the Royal Australian Navy Fleet Air Arm.

The simple, small and light Skyhawk enjoyed a production life of a quarter of a century and is still in front line service with several nations around the world. The Royal Australian Navy received its first of 20 examples in 1967 to replace Sea Venoms and subsequently operated them from the carrier HMAS *Melbourne* until 1982. After withdrawal from service in 1983 the surviving Skyhawks were sold to New Zealand.

The book examines not only the generalities of Australian Skyhawk operations but also the politics behind the end of fixed wing carrier based flying in the Fleet Air Arm, the sale of the aircraft to New Zealand and their subsequent return to Australia in RNZAF colours to support the Australian Defence Forces.

A tight echelon of RAAF F-4E Phantoms over Queensland. The RAAF's use of the Phantom was brief, a temporary measure pending delivery of F-111s.

McDONNELL DOUGLAS
F-4 PHANTOM II

McDONNELL DOUGLAS F-4 PHANTOM II

When one of the great aviation writers describes an aircraft as "the greatest fighter of the postwar era", it is worth taking note. The fighter to which Bill Gunston was referring is the McDonnell Douglas F-4 Phantom II, a genuine multi role combat aircraft which was developed as a carrier borne naval aircraft and not only served with the USA's maritime air arms – the Navy and Marine Corps – but also with the US Air Force which discovered that in the Phantom, the service's nautical 'rivals' had an aircraft superior to anything it could muster.

The result was the first ever purchase of a US Navy fighter by the USAF, purchases of large quantities for firmly land based operations. Of the more than 3,900 Phantoms built for the US services, over 2,600 or two-thirds went to the Air Force.

Eleven other nations also discovered the usefulness of the tough, reliable, superbly performing Phantom. South Korea, Israel, Iran, West Germany, Spain, Greece, Turkey, Japan, Saudi Arabia and Australia all obtained standard or variants of standard models, while a re-engined and largely redesigned version was built for the British Royal Navy and Royal Air Force. The only Phantoms not manufactured at McDonnell's St Louis, Missouri, facility, were 138 for Japan, which were built by Mitsubishi under licence.

The first Phantom flew in May 1958 and the 5,057th and last St Louis built example was handed over in October 1979. The last Japanese Phantom was delivered in May 1981. In the meantime, the Phantom was produced in numerous versions capable of fulfilling a multitude of fighter, bomber, strike, reconnaissance and electronic warfare roles. It is still regarded as a front line type in many parts of the world and will serve for many years to come. It has been successfully deployed in several wars during its operational career – Vietnam, the Middle East, the Gulf – and human nature being what it is, the "greatest fighter of the modern era" will undoubtedly earn more glories before it is finally put to pasture.

Australia's involvement with the Phantom was a relatively minor one in the overall scheme of things and certainly short lived. The Royal Australian Air Force operated the aircraft as much by accident as design and then for less than three years from the second half of 1970.

The circumstances surrounding its entry into RAAF service are fully described in the next chapter, but in summary, 24 F-4E Phantoms were leased from the USAF as an interim measure pending the delayed delivery of a similar number of General Dynamics F-111Cs for Nos 1 and 6 Squadrons at Amberley, Queensland. The Phantoms were delivered in September and October 1970 and the 23 survivors were returned to the USAF in 1972 and 1973. Many were sorry to see them go, but with the arrival of the F-111s, the RAAF felt it had no role for the Phantoms.

Humble Beginnings

When the McDonnell Aircraft Company of St Louis, Missouri, began development of what would become the Phantom II, it was a relatively new organisation without the long traditions of companies like Lockheed, Grumman, Curtiss and Douglas. It would merge with the latter in 1967 to form today's McDonnell Douglas Corporation.

Established in July 1939 by James S McDonnell (or perhaps inevitably, simply 'Mac', as he was universally known), the

The first McDonnell Phantom, the FH-1, delivered in modest numbers to the US Navy from 1946. Its later namesake was built in considerably more substantial quantities. (MDC via Victorian Military Museum)

The supersonic F-101 Voodoo was built in substantial numbers – some 802 from 1954. Some McDonnell family design characteristics which carried over to the Phantom II are evident.

company in its earliest days had a staff of just two (McDonnell and a secretary) and a tiny rented office next to St Louis Municipal Airport, or Lambert Field.

McDonnell's earliest dabble in aircraft design was in 1928 when he produced the Doodlebug, a two seat light aircraft which was entered in the Guggenheim contest for a 'safe' aircraft. Revolutionary in many ways, the Doodlebug's prospects were killed off by the Great Depression, and McDonnell went to work for other aircraft manufacturers, learning the skills which would later allow him to establish his own organisation.

The McDonnell Aircraft Corporation received its first contract for an aircraft of its own design in September 1941 when the USAAF ordered a prototype of the XP-67 twin engined fighter. This extraordinary looking aircraft featured smooth, contoured 'blended' shapes in order to reduce drag, the very heavy armament of six 37mm cannon and experimental powerplants, two Continental XIV-1430-17/90 water cooled inverted V12s of 1,600hp each in very tight cowlings.

The F3H Demon was one of the earlier McDonnell fighters which helped establish the company's reputation as an innovative and reliable manufacturer, despite early problems with the aircraft.

Two views of the prototype XF4H-1 Phantom II, first flown on 27 May 1958. Twenty-three years later the final example was delivered to its customer. (MDC)

The XP-67 flew in January 1944 but was plagued by problems with its powerplants and that, in combination with the fact that its intended role – as a bomber destroyer – was redundant by late 1944, meant the project was cancelled after the completion of just one prototype.

The company had not been idle in the meantime, however. In January 1943 it had received an order for its first jet aircraft from the US Navy – the first Phantom. The first of two XFD-1 Phantom prototypes flew in January 1945 and two months later orders were placed for production aircraft under the new designation FH-1. Sixty examples were delivered to the US Navy and the type made history in July 1946 when it became the first American jet aircraft to land on an aircraft carrier, the USS *Franklin D Roosevelt*. The FH-1 was a conventional straight wing single seat fighter powered by two 1,600lb thrust Westinghouse J30 turbojets and capable of a top speed of 505mph (812km/h).

Modest though the Phantom's success was, it did establish the McDonnell company and allowed it develop a good reputation with the US military. Several successful (and the odd unsuccessful) designs followed in the late 1940s and early 1950s including the F2H Banshee naval fighter (first flight 1947, 892 built), the experimental XF-88 twin jet long range fighter of 1949 which grew into the supersonic F-101 Voodoo for the USAF (1954-1961, 802 built), the F3H Demon naval fighter (1951-1959, 519 built) and the extraordinary XF-85 Goblin 'parasite' fighter of 1948, a tiny jet designed to be carried in the bomb bay of a Convair B-36 bomber and launched when its parent came under threat from enemy aircraft. The Goblin was then retrieved by the B-36, and although successfully tested, was never ordered.

The Demon and the Voodoo, in particular, established McDonnell as an innovative designer and reliable supplier to the US services. Both aircraft incorporated a high degree of technology for their era and featured aerodynamic and structural innovations which enhanced the company's reputation.

It therefore came as something of a disappointment to McDonnell in 1953 when Chance Vought was announced the winner of the competition to supply the US Navy with its first supersonic fighter. Vought's winning design was, of course, the F8U Crusader.

Towards Phantom II

Undaunted, and unwilling to lose the momentum it had built up with the Demon and Voodoo, McDonnell pressed on with a new naval strike fighter design without official support. Like so many inspired aircraft designs, the Phantom began life as a private venture without a customer supporting its early development.

McDonnell called the new aircraft the F3H-G/H Phantom II, and its original design resulted from the company pestering just about anyone connected with the US Navy as to what the service's next generation fighter might be required to do. Questionnaires were left with senior and not so senior Navy officers, people were talked to and listened to and others were invited to visit the factory.

A full size mock up of the new fighter was built in 1954 and appropriate people were invited to inspect it. The mock up revealed an aircraft which had a strong family resemblance to its immediate family predecessors but was also generally similar to the Phantom as it would eventually appear.

McDonnell family characteristics were evident – such as the jet pipes emerging from the truncated lower rear fuselage with the tail assemblies mounted above and behind it on the 'tail cone' – but two major physical characteristics of the definitive Phantom were missing, the anhedral tailplane and dihedral outer wings. The F3H-G/H had flat wings and horizontal tail surfaces.

The mock up showed a large single seater powered by two afterburning 8,000lb thrust Wright J65 turbojets (licence built Armstrong Siddeley Sapphires), four 20mm cannon, radar and provision for no fewer than 11 external stores pylons capable of carrying a wide variety of ordnance befitting the aircraft's multi role aspirations. A speed of Mach 1.5 at altitude was promised.

Perhaps surprisingly, McDonnell's efforts were rewarded in November 1954 when the US Navy signed a letter of intent for two aircraft under the designation AH-1, indicating a primarily attack aircraft rather than a fighter. It should be noted that at the time there was no official requirement for such an aircraft and the move was made as much to keep an innovative design team together and allow them to explore the possibilities, as anything else. It turned out to be an exceptionally wise decision.

The US Navy requested one change to the basic design – replacing the J65s with a pair of the new General Electric J79 turbojets which promised – in its early forms – about 15,000lb thrust with afterburner. The J79 would find its first application in the Lockheed F-104 Starfighter, which first flew in 1954.

The incorporation of the larger and considerably more powerful J79s necessitated substantial redesign of the air inlets and ducts. The inlets were made fully variable in order to extract maximum performance from the aircraft and the Phantom was the first fighter to have this feature.

Work rolled on into 1955 but with little purpose as a letter of intent does not constitute a firm order and there was still no official US Navy requirement for the Phantom.

An F-4B of the US Navy's VF-121 Squadron prepares to launch with other Phantoms and Crusaders in the background. (MDC)

The first major production variant was the F-4B for the US Navy, the first of which was delivered in June 1962. External differences compared with the prototypes and early production versions include the enlarged and raised canopy and the larger nose housing APQ-72 radar. (MDC)

The second major Phantom naval variant was the F-4J, first flown in June 1965. Compared with the F-4B, the 'J' featured more powerful engines, aerodynamics refinements and upgraded avionics. (MDC)

The capabilities of the Phantom could not be ignored by the USAF which ordered the F-4C as its first model. Intended as a 'minimum change' variant of the F-4B, it nevertheless incorporated some important changes to make it compatible with USAF equipment and operations. (MDC)

All that changed in the space of an hour in April 1955 when government officials paid a brief visit to the McDonnell factory in St Louis. They brought with them a detailed specification for a fleet defence fighter capable of performing a mission hitherto unheard of – a 250nm (465km) radius of action combined with two hours combat air patrol (CAP) on station. In that mission, the Phantom had to detect, intercept and destroy unfriendly aircraft.

The following fortnight saw frenzied activity on the Phantom's design: it became a tandem two seater (pilot and a very necessary fire control officer); it lost its internal guns and replaced them with four Raytheon Sparrow radar guided air-to-air missiles mounted in a semi submerged manner under the wide and flat fuselage bottom; all but the centreline stores hardpoint (for a 600US gal/2,271 litre external fuel tank) were initially eliminated, although four underwing points were reinstated when the multi role possibilities of the aircraft were recognised; internal fuel capacity was increased to 2,000US gal (7,570 l) in integral wing and six fuselage bladder tanks; the outer wings (which folded) had their chords increased to give the characteristic 'sawtooth' leading edge and 12 degrees dihedral was applied; the one piece slab 'stabilator' was given firstly 15 degrees and then the definitive 23 degrees anhedral to remove it from the turbulent air created by wing wake at high angles of attack and to counter the rolling effect of the outer wings in yawed flight; and a folding flight refuelling probe was designed for the starboard forward fuselage.

The revisions pleased the US Navy and a decision to proceed with the detail and engineering design was given. What emerged was a great advance in fighter/bomber design, although the Phantom's potential was probably not fully appreciated until it took to the air three years later.

Structurally, the new Phantom's wing comprised a one piece centre section and centre wing structure from wing fold to wing fold with the part that passed through the fuselage comprising a torsion box between the front and main spars which was sealed to form two integral fuel tanks. The wing's trailing edge was a one piece aluminium honeycomb structure while the flaps and ailerons were of conventional all metal construction. The trailing edge flaps were hydraulically operated (three independent systems were fitted) as were the 'blown' leading edge flaps and the airbrakes under each wing.

The fuselage was of conventional all metal semi monocoque construction and the undercarriage (stressed for a descent rate of 22ft/sec at 33,000lb) was hydraulically actuated with single main wheels and double nosewheel. Roll control was provided by conventional ailerons plus upper wing spoilers, the wings were swept back at an angle of 45 degrees and the quickly reinstated hardpoints (under fuselage plus two under each wing) were capable of carrying a load of 16,000lb (7,257kg) between them, including Sidewinder heat seeking and Falcon semi active or heat seeking air-to-air missiles in addition to the Sparrows. A centreline gun pod could also be fitted in lieu of the fuel tank.

The variable engine air intakes and their associated ducting were nothing short of masterpieces of design and the major contributor to the Phantom's exceptional performance. They were "designed for optimum pressure recovery over the full operating envelope of the aircraft, each intake constructed so that it did not swallow the moving boundary layer created by the forward fuselage. This stale air was sheared away by large flat ramps and flow smoothness was assured by suction applied through 12,500 tiny perforations in a movable intake wall section [splitter plate]. The movable ramp could be positioned to achieve best pressure recovery in the intake at any Mach number by an automatic system ..."

If the part of the aeroplane where the air goes in received special attention, so did the bits where the jet efflux leaves the aircraft. The result was not particularly pretty but extremely effective, and this was later proven when the Rolls-Royce Spey powered Phantom for Britain was developed. The Spey turbofans' greater diameters necessitated larger exhausts which upset the previously finely balanced design. The result was that despite having considerably more power, Britain's Phantoms were notably slower.

Orders and Records

In May 1955, McDonnell was given approval to go ahead with the new design and build two prototypes under the designation XF4H-1. As insurance against the failure of the new Phantom, prototypes of the Chance Vought XF8U-3 Crusader III

The RF-4C was an unarmed tactical reconnaissance variant ordered for the USAF in 1963. Three decades later, the type is still an important part of USAF's order of battle. (MDC)

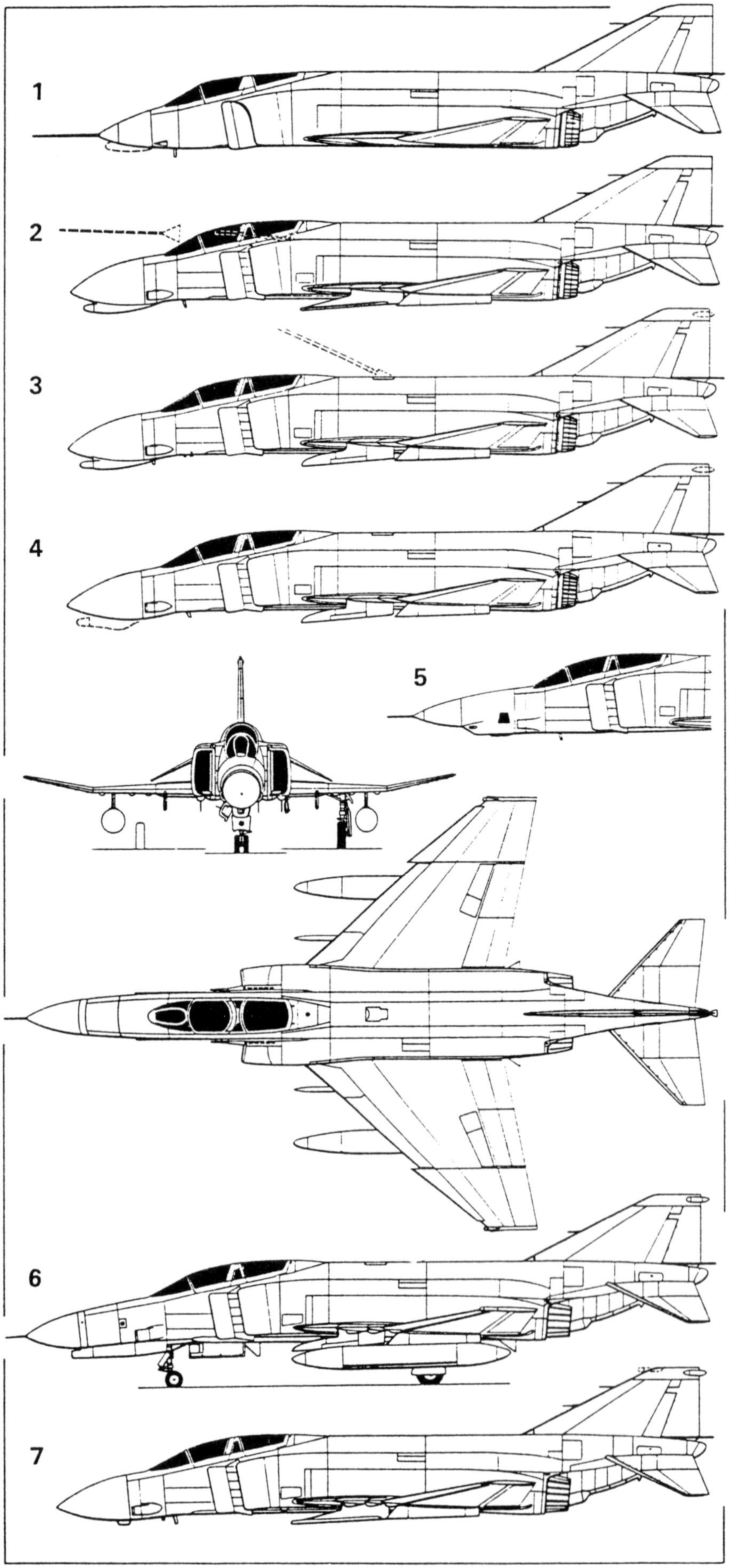

Phantom variants: (1) the first F4H-1 Phantom II with low canopy and the original fixed intake ramps; (2) F-4B with moveable intake ramps and retractable flight refuelling probe; (3) F-4C with 'flying boom' flight refuelling receptacle; (4) F-4D with revised nose, the undernose infra-red detector fitted to early aircraft indicated by dotted lines; (5) lengthened nose of RF-4C; (6) three view drawing of F-4E with revised nose for M61 cannon and APQ-120 radar; (7) F-4J.

were also ordered. A new design with little in common with earlier Crusaders (despite the designation and name), the XF8U-3 was a single engined single seater which carried less than the Phantom, and despite a generally successful test programme, lost the production contract to its McDonnell rival.

Three years after the contract to build the prototypes was awarded, the first McDonnell XF4H-1 Phantom II (BuAer serial number 142259) flew from Lambert Field, St Louis on 27 May 1958 in the hands of company test pilot Robert Little. It was immediately transferred to Edwards Air Force Base for trials. The second aircraft (142260) first flew in October 1958.

Evaluation trials against the Crusader III were completed by late 1958 and in December of that year McDonnell was awarded a contract covering 23 development (including the first two) and 24 production aircraft. These aircraft were designated F4H-1s, although this changed to F-4A when the US forces introduced a new, common designation system in 1962. For convenience, we'll refer to the Phantom by its F-4 designations from now on.

The F-4A Phantom was powered by J79-GE-2 engines each providing 10,350lb dry thrust and 16,150lb with afterburner and were used for test, evaluation and training duties. The first 18 aircraft featured the original 'flat' canopy design but from aircraft number 19, the definitive Phantom raised canopy (and seat line) was fitted. Early aircraft were fitted with 24 inch diameter scanner dish APQ-50 radar, while from number 19 APQ-72 radar with its 32 inch diameter dish was installed, necessitating a larger nose radome. An infra-red seeker was mounted in a small housing under the radome and the physical result was identical to that which would appear on the first major production model, the F-4B.

In the meantime, the Phantom set about proving its prodigious performance with a series of world speed, altitude and time to height records. The first of these was set in early December 1959 when Cdr Lawrence Flint, USN, took the second prototype to 98,577ft (30,046m) in a zoom climb ('Project Top Flight') to set a world absolute altitude record.

No fewer than 12 other class or world records were set by F-4A Phantoms between 1960 and 1962, among them: world absolute speed ('Project Skyburner') 1,606.3mph (2,585km/h) or Mach 2.6 in November 1961 (Col Bob Robinson USMC); 500km closed circuit speed 1,216.76mph (September 1960, Lt-Col Thomas Miller

The final and most numerous new production Phantom variant, the F-4E with solid state APQ-120 radar and M61 cannon under the nose. This example is carrying an EROS collision avoidance unit during test flying by McDonnell Douglas. (MDC)

The most obvious new feature of the F-4E was the undernose 20mm General Electric M61 multi barrel cannon with 640 rounds of ammunition. The F-4E was the first (and only) Phantom with a fixed internal gun. This shot shows some good underside and nose detail.

USMC); 100km closed circuit speed (in effect a constant 3g turn over a period of 2m 40.9sec in this case) 1,390.24mph (September 1960, Cdr J F Davis USN); and possibly the most prestigious of all the speed records – absolute speed over 3 kilometres at a height not exceeding 100 metres or 328 feet. In an aircraft such as the Phantom, such low altitude antics can only be imagined. The Phantom took the record in August 1961 ('Project Sageburner') when Lt Huntington Hardisty USN recorded a speed of 902.769mph (1,452.8km/h) or Mach 1.25 – and all at no more than 175 feet above the ground!

'Project High Jump' covered a series of time to height records set by F-4B Phantoms in 1961-62. Measured from a standing start on the runway, the flights resulted in some impressive figures: to 9,000m (29,520ft) 61.68sec; to 20,000m (65,600ft) 2min 58.50sec and to 30,000m (98,400ft) 6min 11.43m. The latter (flown by Lt Cdr Del Nordberg USN) had its peak trajectory exceed 100,000 feet, or higher than the absolute altitude record set more than two years previously.

The records proved the Phantom's extraordinary performance, but subsequent production models would prove its efficacy in the combat roles for which it was designed.

The Navy's Phantoms

The first major production Phantom for the US Navy and US Marine Corps was the F-4B, externally similar to later F-4As (with APQ-72 radar, infra-red detector and raised canopy) but powered by two J79-GE-8 engines offering 10,900lb dry thrust and 17,000lb with afterburner. Maximum take-off weight was 54,600lb (24,766kg).

The more powerful engines made slightly larger inlets necessary in combination with changed inlet ramp geometry. The first F-4B flew in March 1961, deliveries (to USMC Squadron VMF-314) began in June 1962 and production amounted to 637 between then and the final delivery in January 1967.

A further 12 aircraft were produced as F-4Gs, incorporating an ASW-21 two digital data link. This 'G' model is not to be confused with the later F-4G 'Wild Weasel' radar suppression aircraft which is based on the later F-4E. These early 'Gs' were built in 1963-64 and used by the USMC in Vietnam. They were later converted to standard F-4B configuration.

F-4B conversions included the RF-4B photo-reconnaissance variant for the USMC, using the reconnaissance systems of the USAF's RF-4C. Forty-six conversions were completed from March 1965. The designation DF-4B was given to some Phantoms modified for use as 'mother' aircraft to remotely piloted vehicles (RPVs) and drones, while the QF-4B was an RPV itself which was used for various research, weapons and combat trials. Although controlled from the ground, the nearly 44 Phantoms so converted usually flew with a safety pilot on board.

A more numerous F-4B conversion was the F-4N, the first of 228 flying in February 1973. The idea was to bring older surviving F-4Bs up to a standard similar to that of the F-4J (see below) by rebuilding them (including the incorporation of some new structural parts) and fitting upgraded equipment, systems and avionics.

The Phantom production line at St Louis in early 1978 with the 5,000th example (second from bottom) in final assembly. F-15 Eagles share the production line. (MDC)

This head on view of a late production F-4E shows the extended leading edge 'manoeuvring slats' fitted from 1972 and retrofitted to most surviving earlier F-4Es. (MDC)

The 5,000th Phantom (top) was rolled out in March 1978 and suitably decorated for the occasion. The 5,057th and last St Louis built Phantom was delivered in October 1979. Licence production continued in Japan until 1981. (MDC)

The second major naval Phantom variant was the F-4J, 522 of which were built. The prototype flew in June 1965 and the first production example in May 1966. Production ended in January 1972.

Intended primarily as an interceptor but retaining full ground attack capability, the 'J' introduced several major improvements to its capability. On the engine/airframe side, power was provided by two J79-GE-10 engines rated at 17,900lb thrust with afterburner, while larger main wheels, drooped ailerons and slotted tailplanes were also fitted, the latter pair of changes to help keep approach speeds down in the face of increased weights. The undercarriage was restressed – also to cope with higher weights – and the infra-red detector housing under the nose disappeared.

Internally, the fuel was rearranged into seven fuselage tanks plus the four integral wing tanks, although capacity was slightly reduced. Avionics/weapons systems improvements included the installation of a Westinghouse AN/AWG-10 X-band pulse Doppler fire control system and a new bombing control system.

The F-4Js were subsequently subject to numerous systems/electronics upgrades, but structural refurbishment of 302 aircraft from May 1978 resulted in a new designation, F-4S. Modifications include the replacement of the original wing leading edge flaps with manoeuvring slats, structural strengthening, the fitting of new outer wings and other modifications designed to prolong the aircraft's life.

Britain's Royal Air Force became an F-4J operator in 1982 when 15 second-hand aircraft were purchased to equip a squadron which replaced a previously UK based unit which was sent to defend the Falkland Islands following the skirmish with Argentina in the same year.

Export F-4Es for (top to bottom) Iran, Turkey and Greece. Of the 1,387 F-4Es manufactured at St Louis, 556 were exported. (MDC)

Phantoms for the Air Force

By 1961, the US Air Force had little choice but to take a close look at the Phantom. Its investigations led to the unprecedented action of purchasing an aircraft which had been designed for carrier operations. The USAF found in the Phantom an aircraft which was superior in most respects to anything it had at the time – better radar than any USAF fighter, better payload/range characteristics than any USAF attack aircraft, better serviceability and less maintenance burden than any USAF supersonic fighter and considerable potential for development. Interestingly, the best USAF serviceability/maintenance record among its supersonic fighters was held by McDonnell's own F-101 Voodoo.

In 1962, following evaluation of some F-4B Phantoms, the USAF decided to order the aircraft in large quantities, large enough in fact to re-equip 16 of the service's 23 Tactical Air Command Wings.

The USAF's first Phantom model was at first designated the F-110A but this was quickly changed to F-4C as part of the general realignment of USAF/USN designations in 1962. Externally similar to the Navy's F-4B, the 'C' was a 'minimum change' variant which retained folding wings and an arrestor hook, as did all subsequent versions intended for shore based operations.

Nevertheless, there were some important changes – the addition of a cartridge/pneumatic starting system instead of the Navy's method which required an external hose connection; the replacement of the flight refuelling probe designed for the Navy's probe and drogue system with a receptacle on the top of the fuselage behind the canopy for the USAF's 'flying boom' system; the fitting of larger tyres offering better footprint characteristics and allowing the installation of bigger, anti skid brakes; more complete dual controls in the rear cockpit (at first, both USAF Phantom crew members were pilots but later the 'backseater' weapons systems operators were not necessarily so); the rear instrument panel was lower for a better view; and various items of equipment were modified or upgraded.

The F-4C's avionics were different to its predecessors' and included APQ-100 radar and a new navigation computer in the form of the Litton ASN-48 inertial system. Power was provided by J79-GE-15 engines which gave the same power as the F-4B's -10s but had minor modifications including those necessary for the cartridge starting system.

A Royal Navy Phantom FG.1 (F-4K) of 892 Squadron launches from HMS Ark Royal *in 1972. (MDC)*

The first F-4M Phantom FGR.2 for the Royal Air Force at the time of its maiden flight in February 1967. The RAF retired its Phantoms in 1992. (MDC)

The first F-4C flew in May 1963 and production ended in May 1966 with the delivery of the 583rd example. Some were converted to EF-4C configuration for 'Wild Weasel' electronic reconnaissance and electronic countermeasures (ECM) duties. Thirty-six former USAF F-4Cs were delivered to the Spanish Air Force in 1972.

At the same time the USAF was initially evaluating the Phantom to re-equip its tactical fighter wings, it also looked at an unarmed tactical reconnaissance variant. The result was the RF-4C, ordered in 1963 to replace the McDonnell RF-101C Voodoo.

The first RF-4C flew in August 1963 and the first production model in May 1964. The 'RF' remained in production until the end of 1973 by which time 505 had been built.

The RF-4C retained the powerplants and other mechanical characteristics of the F-4C but featured a lengthened nose housing APQ-99 forward looking radar, a low altitude camera and APQ-102 side looking radar. Most of the remainder of the F-4C's tactical equipment – radar, fire control system and weapons – were removed, providing space for the numerous items of equipment associated with the tactical reconnaissance role: cameras, radar, infra-red/Linescan, various electronic intelligence gathering (ELINT) devices, jammers, electronic countermeasures (ECM) and so on.

The result was an extremely capable tactical reconnaissance aircraft which was able to fly long distance sorties in any weather, and even today the RF-4C is highly regarded as its equipment has been upgraded over the years. In 1992 the USAF still had some 200 RF-4Cs in service.

In March 1964 the go ahead was given for the production of a Phantom variant tailored to the needs of the USAF rather than being a relatively simple modification of a naval type, as was the case with the F-4C.

The result was the F-4D Phantom, 825 of which were manufactured between the prototype's first flight in December 1965 and the final delivery in February 1968. Of that total, 36 were delivered to Iran while 18 went to South Korea in 1969 from USAF stocks.

Powered by J79-GE-15 engines, the F-4D was based on the F-4C airframe but had a substantially updated equipment/avionics fit which included smaller and lighter (due to being partially solid state) APQ-109 radar, an improved sight, weapons release computer and inertial navigation system (INS) along with the capability to carry a wider range of weapons including all the laser guided ('smart') bombs in the USAF's arsenal at the time. Most F-4Ds lacked the distinctive infra-red detector housing under the nose.

The F-4D became the mainstay of USAF Phantom operations in Vietnam and eventually replaced the 'C' model in that conflict. Its period of manufacture coincided with the peak of Phantom production, spurred on by the needs of the war. Of the 5,057 Phantoms built at St Louis, no fewer than 3,037 (or 60 per cent) of them were manufactured in the five year period 1965-69, an average of 607 per annum.

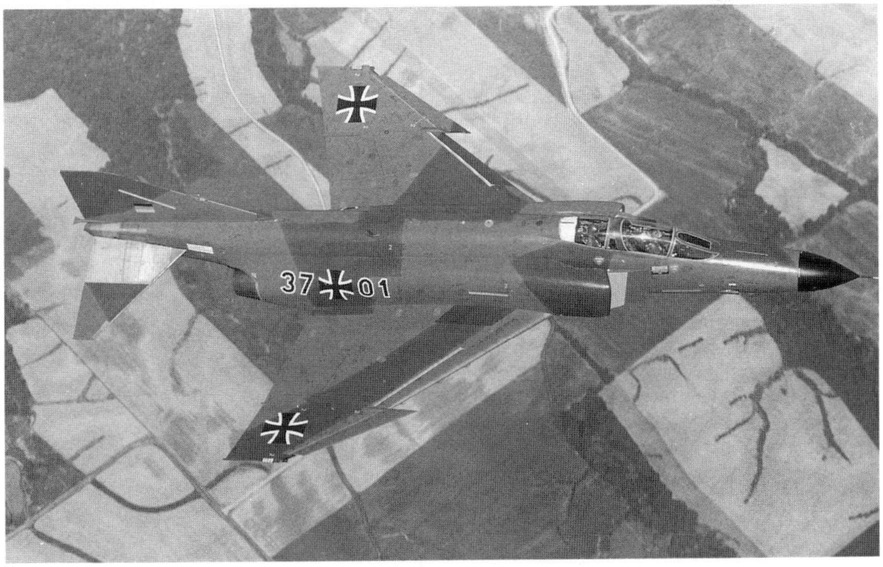

More export Phantoms: (top to bottom) Israeli F-4Es show the effectiveness of their camouflage, German F-4F and Japanese F-4EJ. Japan was the only country outside the USA to build Phantoms, Mitsubishi delivering 138 between 1972 and 1981. (MDC)

The first F-4K Spey Phantom for the Royal Navy, first flown in May 1966 and involving considerable redesign to accommodate the British engines. (MDC)

Annual production for that period was: 1965 – 586; 1966 – 618; 1967 – 759; 1968 – 635; 1969 – 439. Production dropped to 282 in 1970 and averaged about 118 per annum after that.

The Definitive 'E'

The final new production Phantom variant was also produced in the greatest numbers. The YF-4E prototype flew in August 1965, followed by the first production F-4E in June 1967 and deliveries to the USAF three months later. Of the 1,387 F-4Es manufactured at St Louis, 831 went to the USAF and 556 were exported to Israel, Australia, Germany, Greece, Egypt, Iran, Japan, Turkey and South Korea.

The F-4E embodied improvements resulting from lessons learned in combat, the most distinctive of which was the incorporation of a 20mm General Electric M61 multi barrel cannon under the lengthened nose.

American fighter ace, Colonel Robin Olds, once said: "A fighter without a gun ... is like an airplane without a wing". Even though the Phantom was capable of carrying external gun pods, these were far from perfect as they were subject to operational limitations and suffered from shot dispersion caused by the pylons on which they were mounted flexing during firing.

Using missiles only, Phantoms had been successful in numerous air-to-air combats in Vietnam, but many opportunities were missed due to the lack of a fixed gun. The rules of engagement were that positive visual identification of a target had to be made before firing, and that requirement meant the target aircraft was often too close for the Sparrows and Sidewinders to be used.

The gun (with 640 rounds of ammunition) was mounted below the nose, which itself contained a new item of equipment for the F-4E, the more capable, lighter, smaller and completely solid state Westinghouse APQ-120 radar.

McDONNELL DOUGLAS F-4E PHANTOM II

Powerplants: *Two General Electric J79-GE-17A axial flow turbojets each rated at 11,870lb dry thrust and 17,900lb thrust with afterburner; internal fuel capacity 1,855US gal (7,022 l) in seven fuselage bladder tanks and two integral wing tanks; provision for one 600US gal (2,271 l) underfuselage drop tank and two 370US gal (1,400 l) underwing drop tanks; aerial refuelling receptacle in mid upper fuselage.*

Dimensions: *Wing span 38ft 7.5in (11.77m); length 63ft 0in (19.20m); height 16ft 5in (5.00m); wheel track 17ft 9in (5.41m); wing area 530sq ft (49.24sq m); aspect ratio 2.82:1; centre wing dihedral nil, outer wing dihedral 12 deg; sweepback 45 deg.*

Weights: *Basic empty 31,853lb (14,448kg); normal t-o weight 58,000lb (26,309kg), max t-o weight 61,795lb (28,030kg); 'g' limits 7.75 subsonic and 5.93 supersonic at 58,000lb.*

Armament: *One General Electric M61A-1 20mm multi barrel cannon under nose with 640 rounds; four Sparrow radar guided AAMs semi recessed under fuselage; one centreline and four underwing pylons capable of carrying up to 16,000lb (7,257kg) of stores including drop tanks, Sidewinder or Falcon AAMs, conventional and nuclear bombs, mines, fire bombs, cluster bombs, practice bombs, flares, rocket packs, ECM pods, tow targets, cameras pods, spray tanks and Walleye or Shrike air-to-surface missiles.*

Performance: *Max speed (with external stores) 1,432mph (2,306km/h) or Mach 2.12 at 36,000ft, 910mph (1,464km/h) or Mach 1.2 at 1,000ft; max initial climb (clean) 49,800ft (15,180)m/min, service ceiling (clean) 58,750ft (17,907m); max ferry range (full external fuel) 1,612 mls (2,594km) at 580mph (933km/h) at 40,000ft. Refer 'Loading and Performance Data' table for greater detail.*

Other changes included the introduction of a seventh fuselage fuel cell, the slotted stabilator of the F-4J and J79-GE-17 engines, each producing 11,870lb dry thrust and 17,900lb thrust with afterburner. As can be seen from the accompanying table, the F-4E was capable of a wide variety of missions.

The second major change introduced on the F-4E came well into the model's production life but was subsequently retrofitted to earlier aircraft. Leading edge wing slats (often called 'manoeuvring slats') were fitted to new production aircraft from June 1972, which was too late to see service in Vietnam. Again, their fitting was a result of combat experience.

Vietnam proved yet again that dogfights tend to get lower, slower and tighter as they progress and combat against the more nimble MiGs often resulted in Phantom pilots pulling too tight at low altitudes with the inevitable stall/spin into the ground.

The slats improved the Phantom's turning ability considerably and increased the margin of safety. At Mach 0.9 a slatted F-4E could complete a 180 degree turn of 0.7nm radius in 14 seconds, whereas an unslatted example would be 0.4nm behind and 0.2nm outside, a considerable distance. The fitting of slats to the F-4E was not a simple matter and required what was virtually a rebuild of the wings.

Variants of the basic F-4E included the tactical reconnaissance RF-4E for export only (130 built for West Germany, Greece, Iran, Israel, Japan and Turkey) which combined the F-4E basic airframe with the RF-4C equipment fit. The F-4F for Germany (175 built between 1973 and 1976) lacked stabilator slots, some of the classified US electronics, one fuselage fuel cell and Sparrow missile capability but included major airframe components and engines which were manufactured in Germany and shipped to St Louis for final assembly. Earlier proposals to Germany had revolved around a lightweight single seater Phantom called

McDONNELL DOUGLAS F-4E PHANTOM
Loading and Performance Data, Typical Missions

		I Hi-Lo-Hi	II Hi-Lo-Hi	III Lo-Lo-Lo	IV Hi-Lo-Hi	V Hi-Lo-Hi	VI Hi-Lo-Hi IFR	VII CAP	VIII Ferry Range
TAKEOFF LOADING CONDITION		(4) AIM-7E + (1) MK-28 + (2) 370gal tks	(4) AIM-7E + (1) MK-28 + (2) 370gal tks	(4) AIM-7E + (1) MK-28 + (2) 370gal tks	(4) AIM-7E + (6) M-117 + (1) 600 + (2) 370gal tks	(4) AIM-7E + (11) M-117 + (2) 370gal tks	(4) AIM-7E + (1) MK-28 + (2) 370gal tks	(4) AIM-7E + (1) 600 + (2) 370gal tks	Clean + (1) 600 + (2) 370gal tks
TAKEOFF GROSS WEIGHT	lb/*(kg)*	53,814 *(24,410)*	53,814 *(24,410)*	53,814 *(24,410)*	61,629 *(27,954)*	61,795 *(28,030)*	53,814 *(24,410)*	55,991 *(25,398)*	54,171 *(24,490)*
Fuel (JP-4 at 6.5lb/gal (0.78kg/l) Int/Ext	lb/*(kg)*	12,058/4,810 *(5,474/2,183)*	12,058/4,810 *(5,474/2,183)*	12,058/4,810 *(5,474/2,183)*	12,058/8,710 *(5,474/3,954)*	12,058/4,810 *(5,474/2,183)*	12,058/4,810 *(5,474/2,183)*	12,058/8,710 *(5,474/3,954)*	12,058/8,710 *(5,474/3,954)*
Payload – Missiles	lb/*(kg)*	1,820 *(826)*	1,820 *(826)*	1,820 *(826)*	1,820 *(826)*	1,820 *(826)*	1,820 *(826)*	1,820 *(826)*	–
Payload – Bombs	lb/*(kg)*	2,027 *(920)*	2,027 *(920)*	2,027 *(920)*	4,920 *(2,234)*	9,020 *(4,095)*	2,027 *(920)*	–	–
Wing loading	lb/sq ft *(kg/m²)*	102 *(496)*	102 *(496)*	102 *(496)*	116 *(568)*	117 *(569)*	102 *(496)*	106 *(516)*	102 *(498)*
Stall speed (appr pwr, BLC on)	kt/*(km/h)*	148 *(274)*	148 *(274)*	148 *(274)*	158 *(294)*	159 *(294)*	148 *(274)*	151 *(280)*	149 *(275)*
Takeoff ground run at sl[1]	ft	3,180	3,180	3,180	4,360	4,390	3,180	3,490	3,230
Takeoff to clear 50ft[1]	ft	4,490	4,490	4,490	5,840	5,880	4,490	4,850	4,540
Rate of climb at sl[2]	ft/min	9,340	9,340	9,340	6,570	6,170	9,340	8,600	9,240
Rate of climb at sl (one engine out)[1]	ft/min	5,680	5,680	5,680	3,830	3,500	5,680	5,220	5,620
Time: sl to 20,000 ft[2]	min	2.70	2.70	2.70	4.47	5.02	2.70	3.03	2.73
Time: sl to 30,000 ft[2]	min	5.33	5.33	5.33	12.08	12.89[3]	5.33	6.07	5.42
Service ceiling (100ft/min)[2]	ft	36,000	36,000	36,000	29,750	28,100	36,000	34,850	35,850
Service ceiling (one engine out)[1]	ft	32,500	32,500	32,500	23,150	21,300	32,500	30,650	32,350
COMBAT RANGE	nm/*(km)*	–	–	–	–	–	–	–	1,401 *(2,596)*
COMBAT RADIUS	nm/*(km)*	367 *(680)*	359 *(665)*	237 *(439)*	439 *(813)*	266 *(493)*	741 *(1,373)*	250 *(463)*	–
Average speed	kt/*(km/h)*	506 *(938)*	504 *(934)*	345 *(639)*	502 *(930)*	496 *(919)*	504 *(934)*	506 *(938)*	504 *(934)*
Initial cruising altitude	ft	32,950	32,950	sl	29,050	28,100	32,950	31,950	32,800
Final cruising altitude	ft	39,400	39,400	sl	38,750	38,500	39,400	40,400	40,400
TOTAL MISSION TIME	hr	1.57	1.40	1.63	2.02	1.35	3.03	2.21/1.13[4]	2.79
COMBAT LOADING CONDITION		(4) AIM-7E	(4) AIM-7E	(4) AIM-7E	(4) AIM-7E + (2) TER at BL81.50	(4) AIM-7E + (2) TER at BL81.50 + (1) MER at CL	(4) AIM-7E	(4) AIM-7E	Clean
COMBAT WEIGHT	lb/*(kg)*	41,135 *(18,659)*	42,237 *(19,159)*	41,746 *(18,936)*	43,384 *(19,679)*	41,749 *(18,937)*	45,142 *(20,476)*	39,188 *(17,776)*	35,133 *(15,936)*
Combat altitude	ft	sl	sl	sl	sl	sl	sl	40,000	40,400
Combat speed[1/2]	kt *(km/h)*	759/632 *(1,406/1,171)*	759/632 *(1,406/1,171)*	759/632 *(1,406/1,171)*	723/622 *(1,340/1,153)*	703/614 *(1,303/1,138)*	758/632 *(1,405/1,171)*	1,211/– *(2,241/–)*	1,245/568 *(2,307/1,052)*
Combat climb[1/2]	ft/min	41,300/14,100	40,200/13,690	40,700/13,870	36,500/12,250	36,150/12,170	37,650/12,700	15,750/1,460[5]	19,700/2,280[5]
Combat ceiling (500ft/min)[1]	ft	54,800	54,250	54,500	51,200	49,650[6]	52,950	55,700	58,400
Service ceiling (100ft/min)[1/2]	ft	55,100/42,250	54,600/41,700	54,850/41,950	51,700/40,400	50,250/40,550[6]	53,350/40,450	55,950/43,300	58,750/46,050
Service ceiling (one engine out)[1]	ft	40,150	39,600	39,800	37,650	37,300	38,250	41,100	43,750
Max rate of climb at sl[1]	ft/min	41,300	40,200	40,700	36,500	36,150	37,650	43,300	49,800
Max speed at 40,000ft[1]	kt/*(km/h)*	1,221 *(2,262)*	1,219 *(2,259)*	1,220 *(2,260)*	1,159 *(2,148)*	1,108 *(2,053)*	1,216 *(2,253)*	1,223 *(2,266)*	1,245 *(2,307)*
Basic speed at 35,000ft[1]	kt/*(km/h)*	1,221 *(2,262)*	1,220 *(2,260)*	1,221 *(2,262)*	1,158 *(2,146)*	1,106 *(2,049)*	1,217 *(2,255)*	1,222 *(2,264)*	1,238 *(2,294)*
LANDING WEIGHT	lb/*(kg)*	36,831 *(16,721)*	36,831 *(16,721)*	36,831 *(16,721)*	37,821 *(16,716)*	37,944 *(17,227)*	36,831 *(16,721)*	35,134 *(15,950)*	35,134 *(15,950)*
Ground roll at sl	ft	3,680	3,680	3,680	3,760	3,780	3,680	3,520	3,520
Ground roll (parabrake)	ft	3,040	3,040	3,040	3,110	3,120	3,040	2,910	2,910
Total from 50ft	ft	5,590	5,590	5,590	5,670	5,690	5,590	5,430	5,430
Total from 50ft (parabrake)	ft	4,950	4,950	4,950	5,020	5,030	4,950	4,820	4,820

Notes: [1]Maximum power. [2]Military power. [3]Time to military power, combat ceiling (28,750ft). [4]Mission time/CAP time. [5]Supersonic climb schedule. [6]Subsonic ceiling.

F-4E MISSION PROFILES

HI-LO-HI (Missions I, IV and V)
Takeoff with maximum thrust, climb on course with military thrust to optimum cruise altitude, cruise out at long range speeds, descend to sea level (no credit for fuel or distance), search out target for five minutes at military thrust (for missions IV and V, 15 minutes loiter at sea level), expend stores, climb on course with military thrust to optimum cruise altitude, cruise back at long range speeds. Range free allowances include five minutes at normal thrust and one minute at max thrust at sea level static for starting engines and takeoff, five minutes search at sea level at military thrust (15 minutes loiter for missions IV and V); and a reserve of 20 minutes loiter at sea level at speeds for maximum endurance (two engines) plus 5 per cent of initial fuel load.

HI-LO-HI (Mission II)
Takeoff with maximum thrust, climb on course with military thrust to optimum cruise altitude, cruise out at long range speed, descend to sea level (no credit for fuel or distance); at Mach = 0.9 dash 50 naut miles (93km) to target, expend stores; at Mach = 0.9 dash back 50 naut miles (93km) from target, climb on course with military thrust to optimum cruise altitude, cruise back at long range speeds. Range free allowances include five minutes at normal thrust and one minute at max thrust at sea level static for starting engines and takeoff; and a reserve of 20 minutes loiter at sea level at speeds for maximum endurance (two engines) plus 5 per cent of initial fuel load.

LO-LO-LO (Mission III)
Takeoff with maximum thrust, cruise out at sea level at long range speed, loiter 15 minutes at sea level, expend store, cruise back at sea level at long range speed. Range free allowances include five minutes at normal thrust and one minute at max thrust at sea level static for starting engines and takeoff, 15 minutes loiter at sea level; and a reserve of 20 minutes loiter at sea level at speeds for maximum endurance (two engines) plus 5 per cent of initial fuel load.

IFR HI-LO (Mission VI)
Takeoff with maximum thrust, climb on course with military thrust to optimum cruise altitude, cruise out at long range speeds, refuel to capacity at point of no return (no credit for fuel or distance), cruise out at long range speeds, descend to sea level (no credit for fuel or distance), search out target for five minutes at military thrust, expend stores, climb on course with military thrust to optimum cruise altitude, cruise back at long range speeds. Range free allowances as for HI-LO-HI Mission I.

COMBAT AIR PATROL (Mission VII)
Takeoff with maximum thrust, climb on course with military thrust to optimum cruise altitude, cruise out at long range speeds (climb and cruise = 250 naut miles (463km), patrol at max endurance altitude, climb at maximum thrust to acceleration altitude (40,000ft/12,200m), accelerate at maximum thrust at Mach = 1.5 at 40,000ft (12,200m) and remain at this speed and altitude for two minutes, expend missiles, cruise back at long range speed. Range free allowances as for HI-LO-HI Mission II.

FERRY RANGE (Mission VIII)
Takeoff with maximum thrust, climb on course with military thrust to optimum cruise altitude, cruise out at long range speeds to remote base. Range free allowances as for HI-LO-HI Mission II.

All profiles based on external tanks dropped when empty; racks and pylons retained. AIM-7E missiles retained for Missions I-VI; combat radius increased by about 9 per cent if AIM-7Es not carried.

The F-4G 'Wild Weasel' Phantom, an effective suppressor of enemy defences and proving its worth in combat in the 1990s. This example is carrying an AGM-88A high speed anti radiation missile (HARM).

the F-4EF with simplified systems and structures and heavily restricted ordnance capability. It was for Germany's Phantoms that the wing slats were first mooted.

The only country to manufacture and assemble Phantoms outside the USA was Japan, which in addition to those supplied direct in 1971 had 138 F-4EJs built by Mitsubishi and delivered between 1972 and May 1981 when the last new Phantom from any source was handed over to the Japanese Air Self Defence Force.

An important F-4E variant is the F-4G ('Advanced Wild Weasel') electronic countermeasures carriers for the suppression of enemy defences, particularly those associated with air-to-surface missiles. The Wild Weasel concept was developed during the Vietnam War and tried out by small numbers of EF-4C and EF-4D aircraft.

The idea revolves around the aircraft carrying an array of electronic 'black boxes' which will detect, identify, locate and suppress hostile radar. Also carried is the weaponry to destroy it. Although the gun and capability to carry conventional strike weapons is removed, provision is made for the carriage of Sidewinder air-to-air missiles for self defence. While the F-4G can carry out both the 'search' and 'destroy' phases of its mission, it is often used as part of a larger force within which it locates and suppresses enemy radar while other aircraft strike them. Bristling with large numbers of sensors, jammers and other electronic devices, the F-4G proved a success in combat during the Gulf War.

F-4G conversions totalled 116 aircraft, the first of them flying in December 1975 and the last in 1981.

The Spey Phantom

The only production Phantoms not powered by General Electric J79 engines were the 52 Phantom FG.1s (F-4K) and 118 Phantom FGR.2s (F-4M) delivered to the Royal Navy and Royal Air Force between 1968 and 1971. Of the 52 F-4Ks, 24 went to the Royal Navy and the remainder to the RAF, while all of the F-4Ms were delivered to the latter. With the end of fixed wing flying from conventional aircraft carriers by the RN in 1978, the senior service's surviving F-4Ks were transferred to the RAF.

The decision to purchase Phantoms for the RN and RAF was made in 1964 with the abandonment of the Hawker P.1154 supersonic V/STOL project. Although based on the USN's F-4J, the F-4K featured significant differences, not the least of which was the installation of two Rolls-Royce Spey 202/203 turbofans of 20,515lb thrust with afterburner.

This installation required substantial redesign including a new engine bay, wider inlet ducting, a wider fuselage, larger jet nozzles and a reshaped lower rear fuselage. The aircraft had much British equipment and avionics installed (like the change of engines, a political decision to soften the P.1154 cancellation and job losses which followed) and other changes included a hinged radome so the aircraft could fit on RN carrier lifts and a double extendable nosewheel leg which increased its launch angle of attack.

The result was a Phantom which although useful due to the soundness of the basic design, was not without its shortcomings. Compared with its US built counterparts, the Spey Phantom was slower, more expensive, and had an inferior rate of climb, ceiling and range. Part of the problem was the haste with which the British government demanded the project be put together. First orders were placed in 1965 and the prototype YF-4K flew just over a year later in May 1966, a commendable achievement by those involved in the design work.

The first YF-4M followed in February 1967. This variant for the RAF was a true multi role aircraft lacking some of the F-4K's naval features (provision for catapult launch, extended nose leg) but incorporating anti skid brakes and more capable avionics and nav/attack systems, again mainly of British origin. The F-4K and M comprised about 40 per cent (by value) British content and despite their shortcomings served Britain well until retired in 1992.

Into the 1990s, the Phantom continues to usefully serve most of the nations to which examples were delivered. In 1992 the USAF still had more than 600 on strength (F-4Es, F-4Gs and RF-4Cs) although their numbers are steadily declining.

Upgrading and modification programmes will continue to be undertaken, one of the more notable of which is Israel's Phantom 2000 which incorporates structural modifications to extend the life of its F-4Es along with installation of a new advanced and fully integrated avionics suite.

Another Israeli project is the Super Phantom, which revolves around the replacement of the J79 engines with 20,620lb thrust (with afterburner) Pratt & Whitney PW1120 turbojets. A prototype was flown in 1986 and conversions (with optional canard foreplanes) are being offered to other Phantom operators with the promise of significantly enhanced performance.

Phantom: The greatest fighter of the modern era? Quite probably!

RAAF F-4E Phantom '7214 ' (top) on flight test immediately prior to delivery to Australia in September 1970. The Amberley flight line (bottom) with '7210' in the foreground and Canberras visible behind. (MDC/John Bennett)

RAAF Phantoms '7215' and '7217', the latter taxying in after a display at RAAF Amberley. The Phantom was a favourite with airshow crowds during its brief period of Australian service. (Gordon Reid/Peter James)

Phantoms at RAAF Richmond in April 1971 during the period celebrating the RAAF's 50th anniversary. (Rod Farquhar)

Formation takeoff by RAAF Phantoms (top) and a worm's-eye view of another aircraft (bottom) with a 36 Squadron C-130A Hercules in the background. The latter shot shows excellent underside detail. (D MacDonald/Rod Farquhar)

Phantom '0304' (top) photographed in May 1973, one month before returning to the USAF. The RAAF Museum received an ex-USAF F-4E in 1989 (bottom) which was painted in RAAF colours. Note the slatted wing, not a feature of the RAAF's Phantoms. (Gordon Reid/Bruce Chapman)

Scenes at Amberley in June 1973 with the last RAAF Phantoms preparing to return to the USA in company with KC-135 tankers. The first F-111Cs (bottom) had arrived a few days earlier. (Robert Maxey-Fisher)

THE RAAF's PHANTOMS

Australia's brief use of the F-4E Phantom came about as much by accident as design and was unusual in that the 24 aircraft operated by the RAAF were leased from the US Air Force rather than purchased. All were new aircraft from Production Block 43 built in 1970 and therefore lacking the slatted wing of later F-4Es.

The RAAF's Phantoms were delivered in September and October 1970 for service with Nos 1 and 6 Squadrons, the flying components of No 82 (Bomber) Wing at Amberley, Queensland. No 6 Squadron's Phantoms were returned to the USAF in October 1972, followed by No 1 Squadron's aircraft in June 1973.

Background to the Phantom

Phantoms were obtained for the RAAF as a stop-gap pending the oft-delayed delivery of General Dynamics F-111C bombers to Nos 1 and 6 Squadrons. The squadrons had previously operated Canberras, but by 1970 these aircraft had reached the end of their operational lives.

According to the delivery schedule which accompanied the order for 24 F-111Cs in 1963, the first examples of this advanced, capable and ultimately very successful strike aircraft were supposed to arrive in Australia during the course of 1967.

The six year delay while the aircraft's problems were being sorted out meant that the Canberras ran out of time well before the F-111s were delivered, necessitating the introduction of an interim type to keep the RAAF's home based strike squadrons operational.

The third RAAF Canberra squadron – No 2 – flew its aircraft in Vietnam between 1967 and 1971 and kept them in small numbers until retirement in 1982. In the meantime, these Canberras were used for training, target towing and cartographic survey missions.

The delays to the F-111 delivery schedule were of both a technical and administrative nature. The first was caused by the RAAF deciding on the unique 'C' version of the aircraft, a hybrid of the original F-111A and the FB-111A bomber developed for the USAF's Strategic Air Command. This decision pushed the planned delivery dates back to the second half of 1968, still well within the Canberra's estimated useful life.

At this stage, everything seemed reasonably well under control. The first F-111C for the RAAF flew in July 1968 and all 24 aircraft were formally handed over to Australia two months later in a ceremony at General Dynamics' Fort Worth, Texas, facility.

But – within three weeks of the handover – it was decided to delay delivery of the F-111s pending modification and following the 11th crash of the type. An Australian undergoing training, Flt Lt N Pollock, was aboard that aircraft and the RAAF's F-111s were placed into storage.

It was around this time that structural problems associated with the F-111's wing carry through box (the vital structural 'heart' of the aircraft's variable geometry wings) came to a head with cracks appearing in some places along with the discovery of substandard welds due to the impropriety on the part of a subcontractor charged with the manufacture of the box.

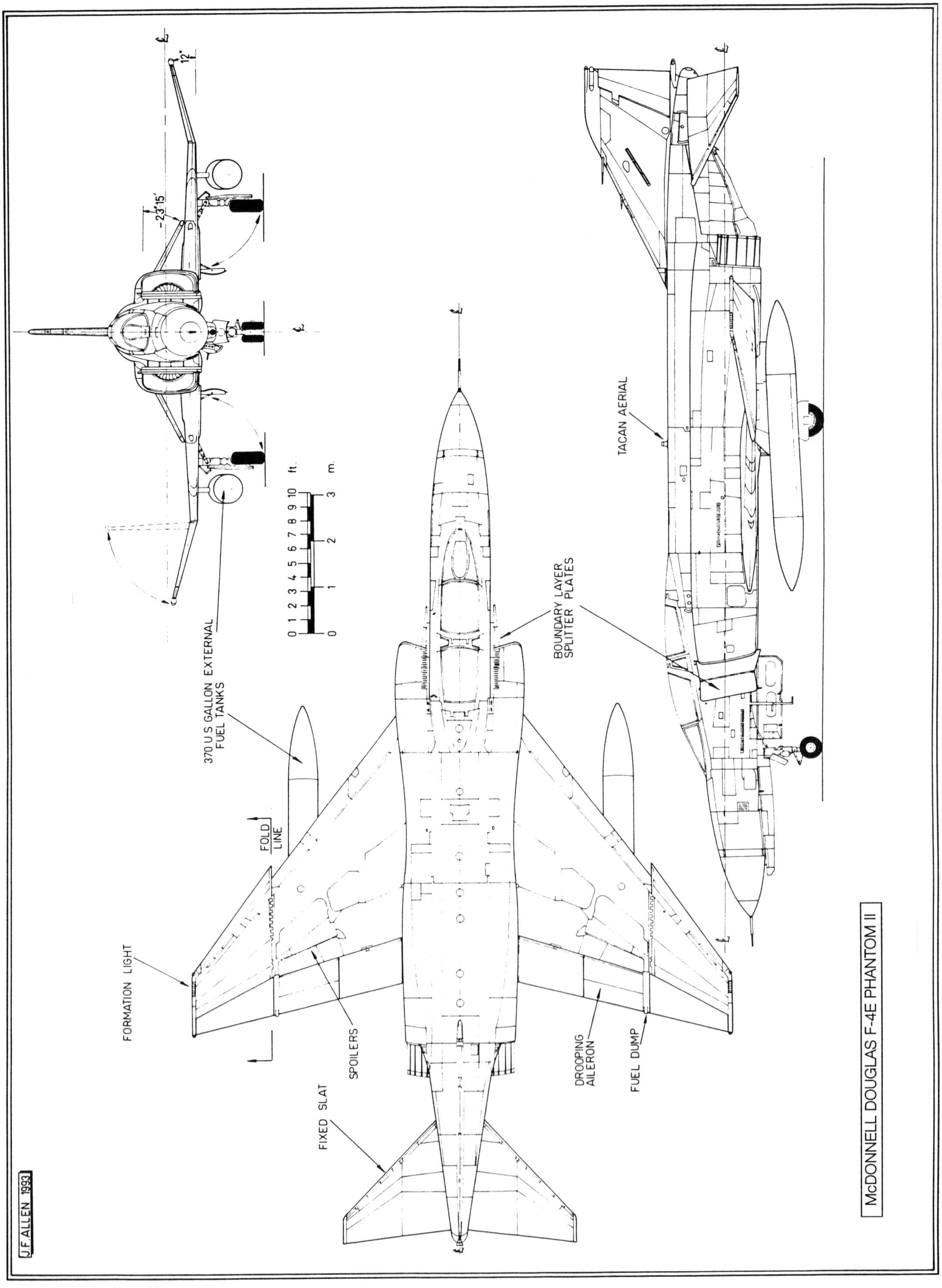
McDONNELL DOUGLAS F-4E PHANTOM II
J.F. ALLEN 1993
370 U.S GALLON EXTERNAL FUEL TANKS
BOUNDARY LAYER SPLITTER PLATES
TACAN AERIAL
FOLD LINE
FORMATION LIGHT
SPOILERS
FIXED SLAT
DROOPING AILERON
FUEL DUMP
-23°15'
12°
0 1 2 3 4 5 6 7 8 9 10 ft.
0 1 2 3 m.

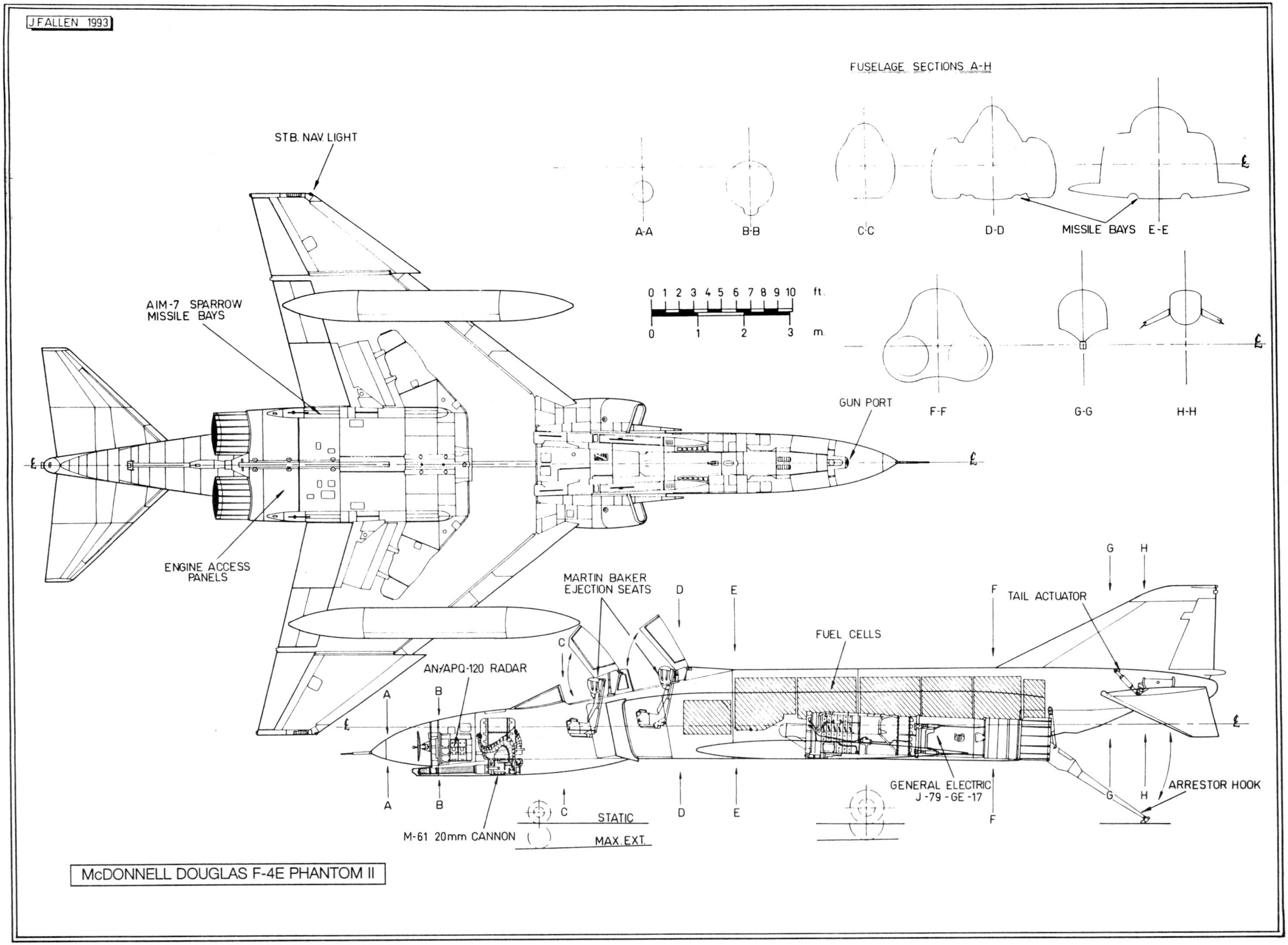

McDONNELL DOUGLAS F-4E PHANTOM II

RAAF Phantom 'A69-69-7209' on approach to Darwin in April 1971. The A69 prefix was never carried by the RAAF's Phantoms and was used purely for administrative purposes. (L Bushell)

The Australian aircraft began to be modified to what was then considered to be an appropriate standard in January 1969, but the discovery of a fatigue crack in a statically tested wing box with 8,000 hours on it in July 1969 once again delayed delivery.

By December 1969 things once more looked as if they were under control and the Australian Defence Minister (Mr Malcolm Fraser) asked the USAF to reactivate the RAAF's stored F-111Cs for early delivery. But more problems followed. In that same month a 15th F-111 crashed when it shed a wing and the entire USAF fleet was grounded for seven months. A massive modification programme was then undertaken, and the already partially dismantled F-111Cs would now lie idle for another two years before reassembly and reactivation was begun. Deliveries to the RAAF finally began in June 1973 and were completed in December.

An Interim Solution

It was the events of late 1969 and early 1970 which prompted the Australian government to seek an interim type to fill the

With the arrival of the General Dynamics F-111C in 1973 (foreground) the RAAF no longer needed its leased Phantoms. Photographed at Amberley the last of them returned to this group of Phantoms in June 1973 just before departure for the USA in the company of USAF KC-135 tankers. (RAAF)

The first RAAF Phantoms arrived at Amberley on 14 September 1970, flown by Australian and American crews. Defence Minister Malcolm Fraser greets Flt Lt Kevin Merrigan (top) while a welcoming speech is made by Air Commodore Doug Kingwell (bottom). Kingwell was the founding CO of 82 Wing when it was formed in August 1944. (RAAF)

The Minister for Air, Senator Tom Drake-Brockman, tries out the cockpit of a Phantom on its arrival in September 1970, assisted by Flt Lt Langlands. (RAAF)

Smiling faces at the Phantom arrival (left to right): Senator Tom Drake-Brockman (Minister for Air), Wng Cdr Roy Frost (CO 82 Wing), Col Hal Hume USAF, Malcolm Fraser (Minister for Defence).

Classics together ... 13 RAAF Phantoms share the ramp with a pair of Royal Air Force Vulcans. (Graeme Gleeson)

RAAF's strike needs while the F-111's problems were being sorted out. At that stage it appeared the aircraft might not be delivered until 1974 and indeed, the abandonment of the programme was a possibility.

Under the direction of AVM C F Read, a mission visited the United States in May 1970 with the intention of organising the lease of 24 suitable aircraft to 'fill the gap' until the F-111 could be delivered. Defence Minister Fraser had set the ball rolling during a visit to the USA the previous month, and the F-4E Phantom was the obvious choice as it was modern, available and proven to be very capable.

The lease deal was announced by Mr Fraser on 22 June 1970 and covered 24 F-4Es which were to be leased at a cost of \$US34m for the first two years and \$US12m per annum after that. These costs were inclusive of spares and ground support equipment and the contract provided for the outright purchase of these and extra Phantoms should the F-111C programme be cancelled.

The RAAF's Phantoms were diverted from USAF orders and delivery began just three months after the contract was announced. As the Phantoms were leased, not purchased, they were given the out of sequence serial number prefix 'A69', this reflecting the USAF serial number prefix and used for stores and accounting purposes only. 'A69' never appeared on the aircraft which retained their USAF serials: 69-0304, 0305, 0306, 0307, 7201, 7202, 7203, 7204, 7205, 7206, 7207, 7208, 7209, 7210, 7211, 7212, 7213, 7214, 7215, 7216, 7217, 7219, 7220 and 7234.

Tail numbers reflecting the last digits of the USAF serials were painted on for identification, the first four aircraft carrying 04 to 07 and the remainder 1 to 17, 19 and 20 and 34.

Preparations

Preparations for the Phantom's entry into RAAF service were quickly organised. In July, the first of 39 members of a USAF Maintenance Training Team who would be based at Amberley arrived. These people would train and advise RAAF maintenance personnel in the early stages of the Phantom programme, conducting courses on Phantom communications, navigation, electronic countermeasures, engines, weapons, electrical systems, instruments and airframes.

In addition, the USAF made available a 20 strong Maintenance Advisory Team, the first of which arrived in Australia in late August. This was supported by the arrival in mid August of the first Phantom ground service equipment (GSE) and early the following month by the arrival of McDonnell Douglas' field service representative, Mr Arthur Hyde.

Personnel from Nos 1 and 6 Squadrons had meanwhile been in the United States since July. Four officers and 66 airmen were undergoing training at MacDill Air Force Base, Florida, the courses covering the Phantom's radar and weapons systems. The first course was completed in early September, allowing its participants to be back in Australia in time for the arrival of the first RAAF Phantoms on 14 September.

Flying training was also undertaken at MacDill, the first conversion course starting at the beginning of August and being completed a month later. It was carried out by the 4530th Tactical Training Squadron.

As can be seen, preparations were by necessity apparently rushed, but nevertheless adequate under the circumstances which revolved around the fact that the Phantoms were leased and not intended to become a permanent part of RAAF strength. This avoided having to establish the complete and complicated infrastructure which would normally accompany the introduction of a sophisticated combat aircraft into service.

Deliveries

The delivery schedule of the RAAF's 24 Phantoms called for four groups of six aircraft to arrive at Amberley on 14 September, 19 September, 26 September and 3 October 1970, respectively. The aircraft were flown from the St Louis factory to George Air Force Base, California and then to Amberley via Hickham AFB Hawaii and Anderson AFB Guam, each flight accompanied by a pair of US Air Force Boeing KC-135 tankers. An RAAF Orion escorted the aircraft on the final leg from Guam, providing weather reconnaissance and search and rescue back up.

The actual delivery dates varied slightly from this, the first five aircraft arriving on 14 September followed by a further six on 19 September, seven on 26 September, four on 3 October and the final two the following day.

The delivery operation was a joint USAF/RAAF effort, the Phantoms crewed by a mix of Australians and Americans. The first batch was led Wng Cdr Roy Frost (temporary CO of 82 Wing) in 69-0306, while 69-7212 was flown by Flt Lt Kevin Merrigan.

The arrival of the first Phantoms at Amberley was accompanied by the usual fanfare, welcoming dignitaries including Defence Minister Fraser, Minister for Air Senator Tom Drake-Brockman and the Chief of Air Staff AM Sir Colin Hannah.

The brief but successful career of the Phantom in RAAF service had begun.

RAAF Phantoms in close formation. The aircraft's Australian career was necessarily brief, although many believed it should have been retained by the RAAF after the F-111s arrived. (RAAF)

PHANTOM IN THE RAAF

Despite the Phantom's short period of service with the RAAF, it filled a large gap in the country's defence capability pending delivery of the 24 General Dynamics F-111Cs ordered in 1963.

As an interim type, the Phantom was not used by the RAAF in the wide variety of roles it was capable of performing but was instead used almost exclusively as a strike aircraft by Nos 1 and 6 Squadrons at Amberley. The idea was to maintain proficiency in that role until the F-111s arrived and training flights and exercise roles were based on that premise.

The Phantoms also provided the two squadrons' ground and technical crews with some expertise on the systems and maintenance of a modern supersonic combat aircraft, and although not as sophisticated as the forthcoming F-111s, it did provide a substantial leap in technology compared with the Canberra.

The Phantoms worked with RAAF Mirage fighters and Royal Australian Navy A-4 Skyhawk strike/fighters during their period of Australian service, taking part in major air defence exercises (including deployments to Darwin for exercises *Top Limit*, *Pepper Pot* and so on) and in naval exercises off the Australian coast. They made good use of the Evans Head bombing and gunnery range on the north coast of New South Wales.

A look at a typical RAAF training sortie 'from the cockpit' appears in the following chapter.

Phantom Squadrons

No 1 Squadron (motto: *Videmus Agamus* – 'We See and We Strike') was the RAAF's first operational unit and can trace its history back to 1916 and the Australian Flying Corps. It saw active service in Egypt and Palestine during the Great War operating a variety of types including the Bristol Scout and Fighter, Handley Page 0/400 and RE 8. The squadron was disbanded in February 1919 and reformed in July 1925 as No 1 Squadron of the by now four years old Royal Australian Air Force.

Between the wars, types operated included the Westland Wapiti, Hawker Demon and Bristol Bulldog, while operations in Malaya (including against the initial Japanese thrust), the Netherlands East Indies and Borneo during World War II saw the use of Lockheed Hudsons, Bristol Beauforts and de Havilland Mosquitos.

Postwar, No 1 Squadron flew Lincolns and Canberras (from 1958) before receiving Phantoms in 1970 and then F-111s three years later. The squadron operated for eight years against the communists in Malaya from 1950, its Lincolns flying 4,000 operational sorties and dropping 33,000,000lb (14,968,800kg) of bombs in the process, the latter figure representing 85 per cent of the tonnage dropped by all aircraft in that conflict.

No 6 Squadron (*Nous Reviendrons* – 'We Shall Return') also had its origins in World War I when as No 30 (Australian Training) Squadron, Royal Flying Corps, it was formed in England in September 1917. Later renamed No 6 (Training) Squadron, Australian Flying Corps, the unit was disbanded in 1919.

The squadron was reformed at Richmond NSW in May 1937 as No 4 (General Reconnaissance) Squadron RAAF initially equipped with Hawker Demons and then Avro Ansons. In January 1939 it was renamed No 6 Squadron and flew Ansons and then Hudsons on coastal patrol duties off New South Wales. Re-equipped with Beauforts, it later took part in operations against Rabaul and other areas of New Britain and New Guinea during the war before being disbanded in October 1945.

No 6 Squadron was reformed at Amberley in February 1948 as part of No 82 (Bomber) Wing and equipped with Lincolns. Canberras were operated from 1956 to 1970 when the Phantoms arrived.

Into Squadron Service

RAAF Phantom flying operations began only three days after the arrival of the first aircraft, the two squadrons initially operating as a single unit under the umbrella of 82 Wing until all the aircraft and crews had arrived from the USA.

Readiness increased from there with various operational milestones being passed during the remainder of 1970 and into 1971. The first night sorties were flown in early October 1970 followed by initial practice bombing sorties in late November, the first defensive tactics sorties (flown

RAAF Phantom 'A69-7212' served with No 6 Squadron and was returned to the USAF in October 1972.

'A69-0305', one of only two RAAF Phantoms which were not converted to F-4G 'Wild Weasel' configuration after their return to the US Air Force.

This shot of 'A69-0305' clearly shows how the USAF serial number was retained on RAAF Phantoms (in standard abbreviated form) with the last two (in this case) digits repeated as a 'buzz' number on the fin. (RAAF)

A quartet of Phantoms on the runway at East Sale with 7234 in front – a rare shot of this aircraft in service with the RAAF. (RAAF)

7206 peels away from a tight formation during a photographic sortie. (RAAF)

Phantom 7219 at East Sale shortly after arrival in Australia. Note the triple ejector release unit under the inner port wing pylon (for three 500lb bombs) and the external fuel tank configuration. (RAAF)

against Mirages) in January 1971, first ground attack sorties in February, and in June high explosive bombing began, some 176 Mk.82 500lb (227kg) 'slick' bombs being dropped over a four day period.

As for the aircraft themselves, there were some delays getting all of the available Phantoms into service due to an early lack of spares, A69-7217, for example, not completing its acceptance test flights until late November 1970 for this reason.

Amberley's support units had plenty to do during the Phantom's short stay in the RAAF. Apart from the accident to A69-7234 (see below) which saw the aircraft out of action for nearly a year before being reflown, No 482 (Maintenance) Squadron and No 3 Aircraft Depot had to install various modifications to the Phantoms' APQ-120 radar (completed June 1971), check all aircraft for defects in the flap emergency system in September 1971 (four Phantoms were found to have loose or broken lines) and following the discovery of a large crack in one of 7234's stabilators during its rebuild, the other Phantoms had to undergo X-ray testing in early 1972. A69-7209 was found to be suffering the same problem.

This same Phantom was the first to have its D2 (300 hours) check performed, in July 1971.

The Phantom's range was put to the test in November 1971 when two aircraft flew non stop from Darwin to Amberley after exercises, a distance of 1,527 nautical miles (2,828km).

The Aircraft Research and Development Unit (ARDU) had only a brief acquaintance with the Phantom, its temporary stay with the RAAF not requiring the usual extensive testing which would normally accompany the introduction of a new type of combat aircraft and the ongoing trials which would be carried out during its service life.

In late June and early July 1971 five Phantom sorties were flown at Amberley for the ARDU in support of arrestor system trials following the accident involving A69-7234 (see below) and a total of 51 trampling passes were made and filmed. These trials were carried out in conjunction with the Aeronautical Research Laboratories (ARL).

3AD's Phantom

There were several accidents and incidents involving the RAAF's Phantoms and at the end of the lease period 23 of the 24 aircraft supplied were returned to the USAF.

The first occurred on 19 October 1970, just over a fortnight after the arrival of the aircraft involved, A69-7234. The following is based on the Court of Enquiry statement into the accident, a story which ended happily with the badly damaged Phantom being rebuilt by No 3 Aircraft Depot at Amberley after many had given the aircraft up as being beyond help.

Phantom '234 departed Amberley in company with three other F-4E aircraft for simulated bombing and strafing attacks on the Evans Head bombing range. The sortie proceeded in a routine fashion until the flight leader, Flt Lt J L Ellis (one of six Australian pilots who had flown USAF Phantoms in

Some good front end detail here as an RAAF crew climbs aboard its Phantom at Amberley in March 1971. Canberras of No 2 Squadron can be seen in the background (RAAF).

Phantoms 7204 and 7210 taking off from Williamtown NSW for a April 1973 air display. (Keith Titmuss)

Vietnam), detected an inflight unserviceability in his aircraft. The left generator had dropped off line and the bus tie-in had failed to function properly.

The crew carried out the checks in accordance with the emergency checklist and when the unserviceability persisted the crew elected to return to base in company with another aircraft from the flight. The crew declared a "Pan" emergency state to ensure that emergency ground rescue services were alerted and activated.

Flt Lt Ellis requested an experienced instructor to be positioned in the Amberley control tower to discuss the situation and possible avenues of action. Four USAF flying instructors, all experienced on Phantoms, went to the tower to lend assistance.

The loss of the left generator and bus tie-in were confirmed, and with these malfunctions two essential services required during landing – brake anti skid and nosewheel steering – were lost.

Twenty of No 82 Wing's Phantoms make an impressive sight lined up at Amberley

Phantom 7204 in flight over southern Queensland.

7202 at an air display during the RAAF's 50th anniversary celebrations in 1971. An RAAF Mirage and Orion are parked behind, while a Royal Canadian Air Force Canadair Argus can just be seen immediately above the Phantom's fuselage. (Michael Melliar-Phelps)

7208 taxying at Amberley. The underwing drop tanks were almost always fitted.

The aftermath of 7234's accident at Amberley in October 1970. An inflight malfunction made an arrested landing necessary, but the wire broke when the aircraft engaged it with the results shown. (Ian Primmer via Sid Finn)

Close up of the damage to the nose of 7234. The aircraft's front fuselage was deemed 'not repairable'. (Ian Primmer via Sid Finn)

Eleven months and 18,750 man hours after its accident, the repaired 7234 emerges from the hangar in September 1972 after what was virtually a complete rebuild. (Ian Primmer via Sid Finn)

Phantom 7213 cavorts above the clouds.

The prevailing 15 knot crosswind on Amberley's long runway would require nosewheel steering and anti skid to keep the aircraft straight during the landing run. After some discussion, the pilot elected to carry out an approach and 'engagement' – that is stopping the aircraft by the use of Amberley's arrestor wires and the Phantom's tail hook.

Flt Lt Ellis had previously carried out two approach and engagements whilst training in the USA as well as a trial engagement at Amberley, so he was familiar with the techniques involved.

While the Phantom circled the area, the Arrestor System Servicing Team assembled its equipment and had ample time to check the whole arrestor system including the tension of the pendant cable, the pre tensioning device and the position of the rubber cable supports. The arrestor system appeared perfectly serviceable.

The aircraft was flown down the final approach "on speed" at about 147 knots and touched down in the centre of the runway about 500 feet short of the arrestor cable. The aircraft's arrestor hook engaged the cable on the runway centreline.

Almost immediately after engagement, the cable on the starboard drum broke and the remaining piece and shock absorber unit were dragged along the runway after the aircraft. The cable slid through the aircraft's arrestor hook until the swaged end fitting on the starboard end of the pendant came to rest against the hook.

At this point the Phantom yawed to starboard, the arrestor hook broke off the aircraft and the starboard arrestor system and remaining cable continued under the aircraft at high speed. The shock absorber flew around the front of the rack under the port wing and then continued back between the port weapon pylon and the port 370 gallon drop tank, rupturing the tank.

The remaining cable wrapped around the port undercarriage strut and broke into pieces as the port wheel ran over the cable.

By this time the aircraft had yawed approximately 60 degrees to starboard and continued to slide along the runway. The pilot tried to straighten the aircraft and had just about succeeded when it left the runway. The port wheel then sank into soft earth and the Phantom yawed violently to port through about 80 degrees.

Then the starboard and nosewheels collapsed and the aircraft came to rest about 2,500 feet (760m) from the end of the runway and 200 feet (60m) off the runway on the starboard side.

Flt Lt Ellis suffered a minor strain in his lower back and his navigator (Flt LT E B J Bolger) was unhurt. The Phantom was in rather worse condition.

The enquiry into the accident found that its basic cause was the fact that the arrestor wire broke when the Phantom engaged it. Causes rejected included the aircraft's arrestor hook, the weather and aircrew fitness and technique – "the aircraft was flown into the arrestor system with considerable skill", the court found.

In summary, the immediate cause of the accident was found to be: "... the fracture of the arresting system cable at the western brake drum, which in turn caused the aircraft to skid out of control until the undercarriage collapsed.

Phantom 7202 of No 1 Squadron taxies out for its solo display at the RAAF Edinburgh Jubilee Air Show in 1971. (Stephen Corbally)

"The cause of the fracture of the cable was a culmination of: having the pickup point of the cable on the drum too close to the flange plate thus making it easier for the cable to whip across the top of the flange; having too much slack cable between the brake drum and the shock absorber, thus enabling the cable to have excessive random movement near the drum; and, the wedging caused by the cable being forced under the outer layers of the coils of cable".

The accident resulted in Amberley's arrestor cables (at both ends of runway 15/33) being declared unserviceable until tested and modified accordingly; the arrestor system being subject to "vigorous trampling trials" before any further practice or operational engagements (undertaken by ARDU and ARL, as mentioned above); RAAF F-4E operations were restricted to daylight sorties until the arrestor system was reinstalled and approved for normal operations; and the forecast maximum crosswind component for daylight Phantom operations was limited to 12 knots.

The CO of No 82 Wing, Wng Cdr R E Frost, further recommended that future planned wire engagements at Amberley be filmed, particularly in the early days of Phantom operations. This would allow assessment of its performance and of pilot technique when engaging the wire.

Phantoms on the ground (top to bottom): 7209 with Sabres in the background; 7207 in the Northern Territory during 1971; and 7216 during the RAAF 50th Anniversary Air Show at Edinburgh SA in April 1971. (Steve Neilson/L Bushell/Stephen Corbally)

The Big Fix

Damaged was sustained to most parts of 234's airframe and components deemed not repairable were the front fuselage, starboard wing and port and nose undercarriage. The aircraft was assessed as having suffered Category 4 damage – repairable at depot or contractor.

A feasibility study of the necessary repair work was carried out in December 1970 and in early February 1971 work began by 3AD in 482 (Maintenance) Squadron's hangar at Amberley. It was carried out by a team of 20 engineering tradesmen headed by Sqdn Ldr C M ('Avro') Anson and supervised by Wnt Off B H Morley.

It was at first thought that the considerable damage would make it necessary for the Phantom to be returned to the USA for repairs, but a USAF team carried out a detailed study of the aircraft in conjunction with 482 Squadron came to the decision that the aircraft could be repaired locally.

Mr Joe Barnes, an American civilian employed by the USAF, acted as repair advisor to the local team.

The job involved a total of 18,750 manhours and most of the special fixtures and handling equipment required to support the repair had to be manufactured in 3AD's workshops. The main parts of the work were replacing both wings, the tailplane and the nose section forward of the cockpit. The achievement was all the greater when the fact that none of the Australian tradesmen involved had any previous experience on the Phantom.

The job was completed in late September 1971 – over 11 months after the accident took place – and '234 was reflown on the 30th of that month to take its place as part

Phantom 7213 climbs out of Amberley for another sortie. Like all but two of the surviving 23 RAAF Phantoms, this one was converted to F-4G Wild Weasel configuration after its return to the USAF.

Some interesting detail in this shot including the triple ejector racks on the inner wing hardpoints, the underwing and centreline external tanks and intake features. (RAAF)

The gathering of the clan? Air and ground crews from No 1 Squadron overwhelm 7210 during Exercise Top Limit at Darwin in May 1972. (Gary Deloas)

of 82 Wing's Phantom force. The job was the first time a repair of such magnitude had been carried out on a Phantom by service personnel anywhere in the world.

'234 lives on in the RAAF, or at least a replica of it does. The real aircraft was returned to its owner in 1972 but the ex USAF F-4E presented to the RAAF Museum in 1989 was painted in RAAF colours as 69-7234 – 3AD's Phantom.

Operational Hazards

There were some other incidents involving RAAF Phantoms, one of them resulting in the loss of an aircraft, No 1 Squadron's A69-7203, and its crew.

On the night of 16 June 1971, Phantoms were exercising near the Evans Head bombing range. At about 8.15pm the exercise had been successfully completed and the aircraft were initiating rejoin procedures when contact with the Phantom was lost. The aircraft was immediately reported overdue and an air search initiated involving RAAF Hercules, Orion, Neptune and Phantom aircraft and civilian aircraft organised by the Department of Civil Aviation.

These aircraft were joined at daybreak by RAAF Iroquois helicopters and an Army Pilatus Porter. On the same morning, an Orion spotted wreckage about 10 miles off Evans Head and a trawler which was operating in the area made its way to the scene to pick up the wreckage for investigation. The wreckage was positively identified as being from the missing Phantom.

Happily, none of the other RAAF Phantom 'moments' ended so tragically. In February 1971 A69-7220 was overstressed in flight (reportedly registering 10g!) resulting in its engines being sent back to the USA for

The old and the new together at Amberley in June 1973. Five days after the arrival of the first F-111s, the last 12 Phantoms began departing for the USA in company with USAF KC-135 tankers. (RAAF)

overhaul and in January 1972 A69-7206 suffered a nosewheel collapse on takeoff which resulted in it being out of service for eight months while repairs were undertaken.

Return to Sender

Two years and one month after their arrival in Australia, the first RAAF Phantoms were returned to the US Air Force. The aircraft were those associated with No 6 Squadron, which flew its last Phantom sortie – an eight aircraft simulated strike and flypast at Amberley – on 4 October 1972.

On 25 October six Phantoms (305, 204, 212, 213, 214 and the legendary 234 – departed Amberley in the company of four USAF KC-135 tankers and flown by American crews. Five more Phantoms (306, 202, 209, 217 and 220) followed in early November.

No 6 Squadron aircrew who were not involved with F-111 familiarisation and training courses in the USA at the time continued to fly Phantoms with No 1 Squadron, which kept its aircraft until June 1973.

On 6 June (five days after the first six F-111s had arrived at Amberley) a similar number of Phantoms departed, followed by four more on 21 June and the final pair the next day. The last official RAAF Phantom flight was recorded on 20 June by 1 Squadron's Sqdn Ldr D N Rogers and navigator Flt Lt L J Cavanagh.

The question of whether or not the RAAF should have kept its Phantoms after the F-111s arrived is one which is still discussed today, two decades after they were returned to the US Air Force.

One of the six RAAF pilots who flew Phantoms in Vietnam was Sqdn Ldr Alan Reed (right), seen here boarding his RF-4C with his 'backseater', Major Melton, USAF. Note the kangaroos painted on the helmets. Their Phantom was called 'Carolina Kangaroo'. Reed flew 100 missions with 12th Tactical Reconnaissance Squadron in 1968, including 10 over North Vietnam. (via AVM Alan Reed)

Certainly, the RAAF wanted to keep them for the close air support role and in 1972 the USAF offered to let Australia purchase the 23 surviving aircraft for what looked to be a generous $A54m for the lot. But when the offer was assessed it was found there would be considerable extra expense involved as spares and ground handling equipment (which had not been purchased under the lease arrangements) would have to be acquired.

According to the then Defence Minister Fairbairn: "If these aircraft [the Phantoms] were brought into the Royal Australian Air Force it would be necessary to disband one Mirage squadron, as well as to recruit and train some 320 pilots and airmen. Therefore, a Phantom squadron could not be fully operational until at least a couple of years' time.

"We should remember that the pilots who are now flying the Phantoms will be transferred to F-111 aircraft when these come into service next year. So for all those reasons the government has decided it should not accept the [USA's] offer.

"At the present time we have in the air superiority and ground attack field in the RAAF and Royal Australian Navy 88 Mirage and 20 Skyhawk aircraft. It was assessed by the RAAF that the F-4E aircraft were superior to these aircraft but they were not markedly superior, only marginally superior ..."

The last statement is an opinion which many would disagree with, but regardless, it meant the end of the Phantom's short service life with the RAAF.

After their return to the USAF, all but two of the 23 surviving RAAF Phantoms were converted to F-4G 'Wild Weasel' configuration. The exceptions were 69-0305 and 69-0307.

Vietnam Phantoms

Australia's involvement with the Phantom did not begin or end with the activities of 82 Wing at Amberley. It in fact started much earlier than that – in 1965 – when the first of six RAAF pilots who would fly USAF Phantoms on operations in Vietnam took his seat in the front of a reconnaissance RF-4C.

Wing Commander Lyall Klaffer was the first Australian fighter pilot to go to Vietnam. An RAAF Korean War veteran (with 77 Squadron), Klaffer was an exchange officer with the USAF, joining the 363rd Tactical Reconnaissance Wing's 16th Tactical Reconnaissance Squadron at Shaw Air Force Base, South Carolina, in March 1965.

When the squadron and its RF-4C Phantoms were sent to Vietnam in October 1965, Klaffer went with them as commanding officer for the first two months of the deployment, playing the leading role in the development of tactics. His was the first RF-4C unit to operate in combat, flying from Tan Son Nhut. Klaffer recorded 51 operational missions in Vietnam before returning to the USA to supervise the training of another RF-4C squadron.

The second Australian Phantom pilot in Vietnam was Sqdn Ldr Alan Reed, a combat instructor pilot with the USAF who hassled his American and Australian superiors into allowing him to go to Vietnam.

With the 12th Tactical Reconnaissance Squadron, Reed flew 100 RF-4C Phantom missions in the period between May and November 1968, including 10 into North Vietnam and often at night using the aircraft's infra-red detectors.

Reed explained some of the operations to author Steve Eather, whose book *Target Charlie* deals with Australian air operations in Vietnam: "Most 'out country' missions flown from Tan Son Nhut were into the lower parts of North Vietnam. I flew most of mine just north of the DMZ (de-militarised zone). On one occasion, however, I flew a mission as far north as Vinh. On another mission into the North I located the crash site of an F-4.

"I flew ten such missions and most were into reasonably well defined areas. Our tactics were to fly at a speed of 600 knots (1,100km/h) at an altitude of 4,500 feet, which was just outside light ground fire range but still under the effective height for launching SAM 3 missiles. At this speed the Phantom used fuel at an alarming rate and we would have to refuel on leaving North Vietnam.

"The RF-4C was totally unarmed and our motto 'Alone, Unarmed and Unafraid' had a certain irony to it ..."

Lyall Klaffer also recorded some flying time over North Vietnam (and Laos) but at that stage neither he nor Reed were restricted by the rule that Australian pilots should not operate over North Vietnam, something which affected the pilots who would later fly USAF Phantoms in the conflict. Also, Klaffer and Reed were flying reconnaissance rather than strike missions and at that time the 'no crossing the border' rule was an unwritten rather than official one and therefore open to interpretation.

Four other Australian pilots flew USAF Phantoms in Vietnam with the 12th Tactical Fighter Wing based initially at Cam Ranh Bay and then Phu Cat. They flew F-4Cs and F-4Ds in the 1969-71 period and were assigned their jobs not so much by individual effort (as had been the case with Klaffer and Reed), but because the Australian government had decided that a small number of pilots should be attached to USAF combat units in order to gain experience.

Sqdn Ldr Hans Roser and Flt Lt John Ellis were the first pair, arriving in Vietnam in October 1969 after completing an F-4C conversion course in the USA. The main job of the 12th TFW was to provide close air support for US, Vietnamese, Korean, Thai and Australian troops on the ground. Unlike their American counterparts in the wing at the time, the Australians were not permitted to fly missions into North Vietnam, although they did fly escort missions for B-52 bombers making raids into Laos.

Flt Lt Ellis was awarded the Distinguished Flying Cross for his actions during the North Vietnamese Army's Pleiku offensive in April and May 1970. His DFC citation read: "On 24 April 1970 he and his wingman were scrambled for a rendezvous with an FAC (Forward Air Control) over Dak Seang Camp. Friendly forces were only 50 metres from the position to be attacked. As the friendly position was receiving extremely heavy mortar and recoilless rifle fire, Flt Lt Ellis led his wing man through eight low level bombing attacks, followed by strafing attacks.

"During every attack both aircraft received small arms battle damage, but five enemy soldiers were killed and a measure of relief was granted to the allied position. Later that evening, Flt Lt Ellis carried out a number of attacks on the same position despite bad weather and failing light".

Both Ellis and Roser returned to Australia in September 1970, Roser's log book showing he flew 165 missions and 230 operational hours.

The last two Australian Vietnam Phantom pilots were Sqdn Ldr Ian Whisker and Flt Lt Lindsay Naylor (who was also awarded a DFC), both of whom were assigned to the 12th TFW in late 1970 to replace Roser and Ellis.

Whisker and Naylor were faced with the same ban on flying into North Vietnam as had been their predecessors, but this time the situation created problems within their units.

There were occasions when the Australians were listed to attack troops or vehicles which were situated just south of the border but they had to abort at short notice because the targets had moved north and were therefore 'out of bounds'. This created embarrassment for the Australians and some resentment on the part of the Americans with whom they had to serve. Some Americans felt the Australians were being 'looked after' by flying against lower risk targets in the south, while they had to take what were perceived as greater risks by crossing the border.

Despite this, there were still some occasions when the Australians hit their targets on the 'wrong' side of the border. Sometimes, the pilots were already close to attacking when the forward air controllers advised the targets had crossed the ill defined border, but in these cases the attacks went ahead as to pull out then would have been laughable. As it was, the squadrons' plans for a particular strike were sometimes subject to late change when it was known the target had moved north just before takeoff.

The situation was clearly unworkable and it was one which had come about for political rather than military reasons – something which sums up the conduct of the Vietnam war generally, some would say.

The F-4C, mount of several of Australia's Vietnam Phantom pilots. (Flt Lt C Chambers)

More shots of RAAF Phantoms departing Amberley for the USA in June 1973. The last official RAAF Phantom flight was flown by No 1 Squadron on 20 June. (RAAF)

FLYING THE PHANTOM

Roy Phillips is a pilot with a vast amount of experience on fast jets. Currently part of Qantas Technical Aircrew flying Boeing 747-400s around the world, he previously spent 20 years with the RAAF specialising in fast jets before retiring with the rank of Wing Commander.

During his RAAF career Roy flew Sabres, Mirages, Phantoms and finally F-111s, logging over 3,300 hours on these four types as well as being involved in the tactical fighter evaluation programme which resulted in the F/A-18 Hornet being selected to replace the Mirage.

With Roy's help, we now take you into the cockpit of the F-4E Phantom to experience an impression of what it was like to fly the aircraft, both in general terms and in the course of a training sortie.

★ ★ ★

For most of the pilots approaching the Phantom for the first time, first impressions would have been the same – the size of the aircraft and its distinctive appearance. It was an aggressive looking aeroplane, very businesslike, but with pieces of the airframe seemingly pointed in wrong or unusual directions. With its long droopy nose (in the case of the F-4E), its dihedral outer wings and heavily anhedral stabilators, it really didn't look as if it should fly at all!

For those who had previously flown the Mirage III, the Phantom offered a marked comparison – it looked like a tank and the cockpit was incredibly spacious and well laid out, in contrast to many European types. The way the controls and switches were laid out were typically American – uncluttered and well thought out.

Visibility over the nose and to the side of the aircraft was excellent but rearward visibility was poor, a fact exacerbated by the front seat being considerably lower than the rear.

Again in comparison to the Mirage, the Phantom's engine and airframe limitations were minimal. The General Electric J79 engines were incredibly reliable and required little or no special handling during start, in flight, at high 'g' or high angles of attack and at supersonic speeds. Engine handling limits were encountered so infrequently, one gradually forgot what the limits were. It was almost impossible to generate compressor stall as the engines were extremely flexible. The throttles could be slammed from idle power to full afterburner and back to idle again without problem, a result of the quality of the engine itself and the aerodynamics which fed it with air. It was a brilliant matching of airframe and engine.

Flying the Phantom was akin to driving a twin turbocharged Mack truck. Brute force alone pushed the unconventional and very 'draggy' airframe through the air, and not with consummate ease. Engine thrust against airframe drag was always apparent on this aircraft; close the throttles at high speed and the aircraft just 'stopped', throwing the pilot against his straps. There was little need to deploy the speed brakes.

Yet, the Phantom could be accelerated to Mach 2.0 with ease and with none of the airframe creaks, cracks and groans which typified other aircraft which were supposedly capable of that speed.

A level acceleration to 600 knots (indicated) at 35-37,000 feet, raise the nose and Mach 2.0 would appear fairly quickly at 45,000 feet with little fuss from the engines or airframe.

Closing the throttles from maximum afterburner to military power and then to idle power produced a spectacular deceleration. It was in this region of flight that the F-111 was (or is) prone to compressor stall, a shuddering thump to the airframe followed by a sheet of flame from *out* of the intake, the result of disturbed airflow into the engines.

At higher supersonic speeds the Phantom possessed considerable kinetic energy, and in combat it is essential to preserve this energy. The Phantom could be manhandled with little fear of inducing any form of compressor stall and resultant loss of energy.

At the other end of the scale, such as in the circuit, the Phantom was a delight to fly, as befits an aircraft originally designed for carrier operations. From an 'initial' of 450 knots and a 4g pitch, the aircraft would arrive downwind at 1,500 feet AGL and 270KIAS (knots indicated airspeed). Rudder would be used to roll out of the turn onto downwind and onto the base leg because of the Phantom's well known adverse yaw characteristics at high angles of attack – high 'g' at high speed or steep turns at low speed.

At 150KIAS, gear down and flap down, check hydraulic pressures and warning lights, gear indicating down and gear handle lights in both cockpits extinguished. Roll onto base using rudder and looking for the previously calculated airspeed and as both the leading and trailing edge flaps are 'blown' by high temperature/high pressure engine bleed air, a special check was required to ensure there was no BLC (boundary layer control) malfunction warning. BLC improved the low speed performance of the Phantom.

Establish on finals with a descent rate of 700ft/min, cross the threshold, wait for

Rotate! An RAAF Phantom departs on another training sortie. (RAAF)

ground effect to flatten the approach slightly and touchdown. Drag 'chute deployed, nosewheel steering engaged below 70KIAS and brakes as required.

Any significant crosswind would weathercock the Phantom into wind, so the pilot needed to be ready to jettison the 'chute should control prove difficult.

Ground Attack Mission

A typical RAAF Phantom sortie might comprise a low level navigation exercise followed by dive bombing onto the Evans Head range and then gunnery either against a target at Evans Head or against a 'splash' target towed by an RAN vessel. The latter was regarded as great 'sport' by the Phantom crews.

Start, taxi and takeoff were usually briefed as SOP – Standard Operating Procedures – for a four ship formation. Callsigns such as 'Buckshot Carbine' or 'Buckshot Cannon' were used. Engine start was via a ground cart providing pneumatic air for engine rotation and although the start temperature limit of 750 degrees was carefully observed it was rarely if ever encountered.

Fuel flow, idle RPM, boost pump pressures, hydraulic pressures, oil pressures and normal generator operation were checked in sequence after engine start. The formation was then checked in to make sure all aircraft were serviceable and taxi out as a four ship was via one hundred yards centreline spacing.

The aircraft were lined up on the runway as two slightly offset pairs so as to prevent the jet efflux of the front pair from going straight down the intakes of the rear two.

Prior to takeoff, the stability augmentation switches were engaged, flight controls and stabilator trim were checked, ejection safety guards were lowered, harness, leg restraints, fuel quantity checked, canopy checked, closed and locked, trailing edge flaps set to half (and leading edge flaps extended) and takeoff was begun.

Rudder became effective at 70 knots and nosewheel steering was disengaged at that point. Aft stick at nosewheel liftoff for a 10-12deg pitch attitude and fly off at approximately 160-165 knots. The gear and flaps were retracted before 250 knots and from there the Phantom accelerated

Turning hard, a Phantom dives in for what might be a gunnery run, judging by the lack of underwing stores.

A typical four ship formation lines up for takeoff, the two rows of aircraft staggered to avoid jet efflux problems for the rear pair. (RAAF)

very quickly with full afterburner engaged. Acceleration was generally to 400 knots into Mach 0.9, the usual climb speed.

Very low level formation navigation was quite a demanding exercise in the Phantom. Those flying wing man or as an element lead during flight could find it difficult to stay in position during tight turns and at the same time maintain visual contact with the lead as well as keeping an eye on the surrounding scenery.

A lot of RAAF low level navigation was conducted around the New England area around the hills and into the valleys, and to fly in the combat position, 600 feet away from your leader in a 30 degree cone around his tail was quite difficult in tight turns as the tendency was to lose sight of him under the nose. Of course, maintaining visual contact under those circumstances was of paramount importance.

The navigators in the rear seat were mainly responsible for the low level track and timekeeping as well as providing another set of eyes to look out for natural and man made obstacles such as crop-dusting aircraft which were often around that area.

General low level navigation was conducted at 480 knots with an acceleration to 540 knots when approaching the target area. The navigation exercises were generally conducted as a four ship formation flying a variety of different tactical formations depending on the nature of the exercise.

The lead aircraft would be responsible for the overall navigation and operation of the formation which would invariably end up just to the south or north of Evans Head prior to weapons delivery practice.

The leader would then call the formation into low level battle manoeuvring formation then perhaps into a finger four, line astern and lastly into an echelon as the Phantoms ran in to the range at low level and at weapons delivery speed, about 500 knots. Each pilot had to make sure his rudder trim in particular was set as this had a significant impact on weapons accuracy.

The aircraft would then pitch up into the weapons pattern to begin work, each aircraft spaced by about five seconds so as to allow four of them to operate within a basically rectangular weapons pattern around the target.

The lead Phantom would set the pattern, the number three man was responsible for flying opposite him in the pattern and numbers two and four filled in the gaps.

The aircraft could carry a good load on its external hardpoints, but 12 Mk.82 500-pounders was a practical load for the F-4E. A variety of bombing modes was available via the weapons systems computer – loft, timed offset, timed lead, timed level, target find, dive toss, dive lay, lay down or offset.

The dive toss mode was used most frequently by the RAAF as the technique provided most flexibility in a tactical environment. Practice in this technique was usually conducted on a daily basis at Evans Head.

In preparation for an attack, the aircraft would be climbed to about 9,500 or 10,000 feet and then rolled into a 45 degree dive at 350 knots. Once established in the dive (and not before), the armament master switch was selected 'on'. There were several instances of pilots turning the switch on early and in one case a Mk.82 bomb was thrown onto the MZ airfield when the aircraft turned onto base because the armament switch was turned on when it shouldn't have been.

Once in the dive, the 'backseater' would boresight the radar and lock onto the ground returns immediately in front of the aircraft's longitudinal axis. The weapons computer in the Phantom was then in receipt of slant range from the radar as well as drift rates, aircraft velocity and dive angle information from the inertial navigation system, allowing the pilot to drift the gunsight 'pipper' towards the target in the normal manner and his computer would generate a bomb release signal once the pickle button on the control column had been depressed.

Weapons release would then occur at 5,500-6,000 feet and 480KIAS and the recovery had to be made clear of the weapons fragment danger area. On some occasions, aircraft were struck by fragments from the exploding bombs, in particular the 500-pounders. Usually, 11.5lb practice smoke bombs were dropped on the range, released from the SUU-20 pod carried on one of the inboard weapons stations.

The fall of the weapon was plotted by the range ground party and the range safety officer would 'call the bomb', for example "25 feet five o'clock", or "20 feet eleven o'clock" ... and occasionally "200 feet, six o'clock!" Each pilot made several passes.

Some Good Sport

Having completed the bombing phase of the exercise, the aircraft would then recover from the last bombing pass to the lower altitude of 4-4,500 feet to begin the gunnery phase.

The gunnery pattern was flown from about 4,000 feet at a firing speed of 480 knots against what were called 'rag targets', banners about 30 feet square or in

Two Phantom aircrew from No 6 Squadron plan their next sortie. (RAAF)

A look around the front cockpit of an F-4E, in fact the example presented to the RAAF Museum in 1989 by the USAF. Top to bottom: instrument panel/right hand side; control column/lower instrument panel; left hand console/throttles. (Bruce Chapman)

the later phases of Phantom operations, acoustic scorers where an acoustic device counted the number of rounds which entered the target area, a fan shaped area generated by the device. The latter gave instant results, of course.

Air-to-ground gunnery was colloquially referred to as 'The Sport of Kings' and gunnery was generally regarded as being a lot of fun. the Vulcan rotary cannon on the Phantom was an absolutely outstanding weapon capable of firing at either 4,000 or 6,000 rounds per minute, selectable by the pilot.

Each gunnery pass was no more than just a squirt, one quick squeeze of the trigger during which perhaps 50-70 rounds were loosed off. Because the cannon rotated so quickly, a number of rounds were lost because of overrun as the cannon wound down once the trigger had been released.

Firing was begun from a 15 degree dive at a slant range of from 1,800 feet into 1,200 feet, allowing for wind effect by placing the sight pipper to the left or right of the target, as appropriate. The 1,200 feet minimum firing range was for safety reasons due to the effects of ricochetting bits of 20mm cannon rounds (the fragmentation envelope) causing danger to the aircraft on the far side of the target during the 4.5g recovery. Several aircraft were damaged this way.

If an aircraft had not finished firing by the time it had reached 1,200 feet slant range, the Range Safety Officer would declare "foul", and if the same offence was called twice, the aircraft involved would have to set its armament switches to 'safe' and return to base.

Splash gunnery would involve a rendezvous with a Navy destroyer or frigate which was towing the target on the end of a long cable nearly a mile long. A four ship Phantom formation would attack the target, flying a rectangular pattern the axis of which was at 45 degrees to the track of the ship, with the ship on the inside of the pattern so the Range Safety Officer aboard the ship could keep all aircraft in sight. In this case there was no scoring, but the fall of shot was easy to see as the rounds hit the water.

As an exercise in sheer good fun, gunnery against a splash target was hard to beat! The Phantom's cannon was of the 'point harmonisation' type where all the rounds ended up grouped very closely – in a figure of eight pattern in this case – with potentially devastating results for the target.

Each aircraft would perform five or six passes against the splash target, the crews doing their best to separate the target from its cable and at the end of it all rejoin for either a formation or single ship flypast for the benefit of the Navy, the done thing, a form of acknowledgement or thanks.

Phantom pilots can all recall some startled sailors looking *down* at the aircraft as they rocketed past the deck at 550 or more knots and dot feet just to say "thanks boys, it was good fun, we'll see you next time". This was of course technically illegal but it had been going on for years, everybody did it and it was just great fun. Low and fast always is!

Characteristics

Some of the Phantom's flying characteristics are worthy of note. Because the aircraft was a bit of a novelty in the RAAF, service officers were keen to go for a fly in it and many did. During these flights the Phantom's handling would be demonstrated, allowing the pilots to perform some manoeuvres which were outside those flown during regular bombing or gunnery sorties.

A demonstration flight would begin with the all important emergency and ejection briefing, although as most of the passengers were generally inexperienced in flying or at least with fast jets, it was clear that most of it was going straight over their heads despite their nodding at the appropriate times! The Phantom's command ejection system (where the pilot could initiate dual or single ejections) would obviously come into play in these circumstances should it have been needed, which it wasn't.

The demonstration takeoff (in a clean and light aircraft) would comprise taking off, cleaning up, holding the aircraft down until 600 knots was indicated and then pulling the aircraft vertical. This has been accurately described as being like 'riding on a rocket', the aircraft shooting straight up with Amberley's runway rapidly diminishing in size and retreating into the middle distance beneath the tail.

The Phantom would be taken up to 15-20,000 feet, levelled out and then have its handling qualities demonstrated. The Phantom's flight controls weren't as responsive as some aircraft (like the Mirage) – the aircraft looked a bit like a truck and it flew like one, albeit one with an enormous amount of power.

The sensitivity of the aircraft in pitch in particular was rather unusual and this was exacerbated by the thrust line of the engines. Selection of the afterburners at speeds from about 500 knots would result in a series of 'kicks' in the backside as the throttles were advanced. Quite often, that 'kick' would produce a nose up response from the aircraft which could be corrected by forward stick, but because of the relative lack of sensitivity in pitch, it was very easy to create a pilot induced oscillation, in which the pilot's control inputs got out of sync with what the Phantom was doing and eventually the pilot could add to the problem.

This happened to RAAF Phantoms on a couple of occasions, aircraft developing pilot induced oscillations, resulting in maximum 'g' being put on the airframe as a loss of control situation was approached. Once pilots got used to this characteristic it was fairly easy to contain, but it was something which kept them on their toes.

While it could be said the Phantom's flight control system lacked a certain *finesse*, in its normal operating envelope – 350 to 450 knots with a heavy load on board – these sorts of problems wouldn't be encountered and it was matter of hauling the aircraft around the sky as a tactical fighter bomber, a role in which it excelled.

The Phantom suffered significantly from adverse yaw – the tendency to turn the opposite way to that which was intended as it was rolled into a turn, particularly at higher 'g' loadings. This characteristic was the result of the combination of control deflections and the basic aerodynamics of the aeroplane. After all, such an unusual shape was bound to have some strange characteristics!

To help counteract this, Phantoms were fitted with an aileron-rudder interconnect which was particularly useful at low speeds or high angles of attack. It helped in the circuit, for example, but at altitude at higher airspeeds and angles of attack there was insufficient rudder authority to overcome the adverse yaw characteristic. The result was that in some circumstances the rudder was more effective in turning the aircraft than were the ailerons, demonstrating at once how effective the rudder was at high angles of attack and how potentially dangerously ineffective the ailerons could be under certain circumstances.

But overall, the Phantom was quite safe provided the control inputs and general operating techniques were appropriate for the prevailing circumstances, something which applies to any aeroplane.

Another interesting demonstration was to pull the nose up and get the Phantom back to 80 knots indicated airspeed, very slow for such a large aeroplane and certainly extraordinary when compared with something like the Mirage.

At 80 knots the control column could be pushed full forward which would deflect the large stabilator, then the pilot could feed in full aileron and roll the aeroplane with few dramas provided the angle of attack was kept within a certain range regardless of the aircraft's attitude with manipulation of the stick forwards and backwards, an extraordinary sight in such a beast. This could be kept up *ad infinitum* with full military power applied, the Phantom gradually accelerating as the manoeuvre progressed.

The drag 'chute was not used just to slow the aeroplane after landing but also as an emergency 'straightener upper' should it depart controlled flight during aerobatics and centralising the controls failed to bring it back. The 'chute would bring the Phantom out of its gyrations and then of course rip away as speed increased to about 200 knots.

In summary, the Phantom was a fabulous aeroplane. Equally effective in a number of roles, it was perhaps unfortunate that the RAAF didn't exploit its capabilities as a fighter, but for a pilot who flew the Mirage and F-111 as well, the Phantom was the best workhorse of the lot – reliable and strong with very few limits and when operated in the normal way, delightful to fly. Pilots just had to be careful at high speeds, high 'g' and high angles of attack and it could bite if provoked.

Definitely the most useful fighter/bomber ever operated by the RAAF – without peer, in fact!

A rare flying shot of '7234', the Phantom which was badly damaged shortly after arrival in Australia and subsequently repaired over a lengthy period (refer previous chapter).

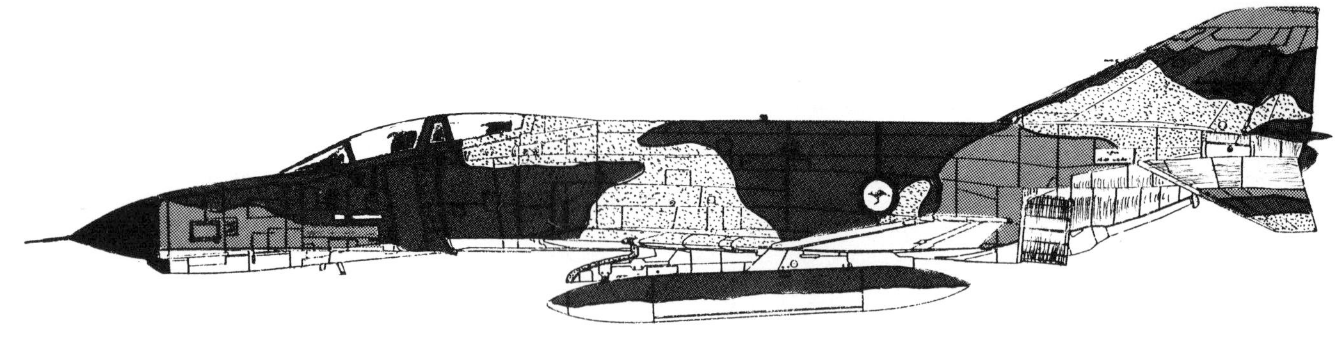

F-4E Phantom II 69-0304, 1970
Standard USAF SE Asia camouflage scheme of dark green, medium green and tan upper surface, with light grey undersurface. Black serial (90304) on fin. Black nose and red turbine warning stripe encircling mid-upper fuselage.

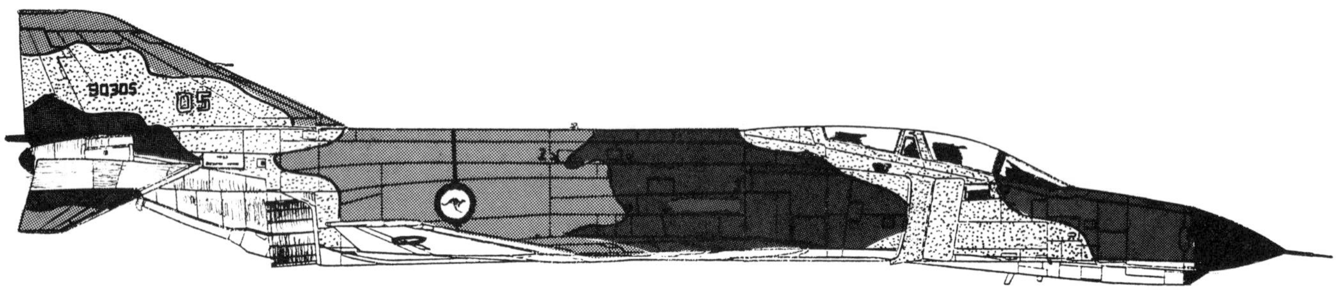

F-4E Phantom II 69-0305, 1971
Standard scheme. Black serial (90305) on fin. The last two digits of the serial are repeated in white. This view shows camouflage pattern for the right side.

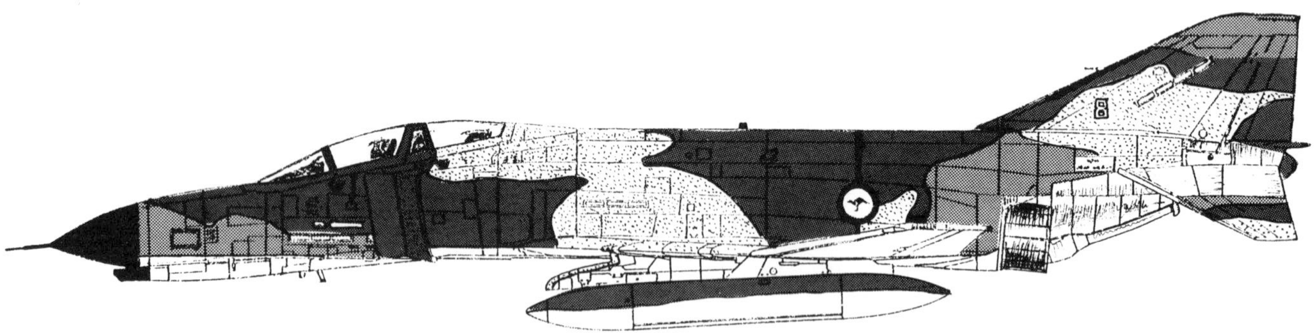

F-4E Phantom II 69-0308, 1972
Standard scheme, with only the last digit of the serial being applied to the fin. This aircraft has been updated with the extended cannon fairing under the nose and the addition of electro-luminescent formation lights on the fuselage, fin and wing tips.

Dark Green (FS 34079)

Medium Green (FS 34102)

Tan (FS 30219)

Undersurface Light Grey (FS 36622)

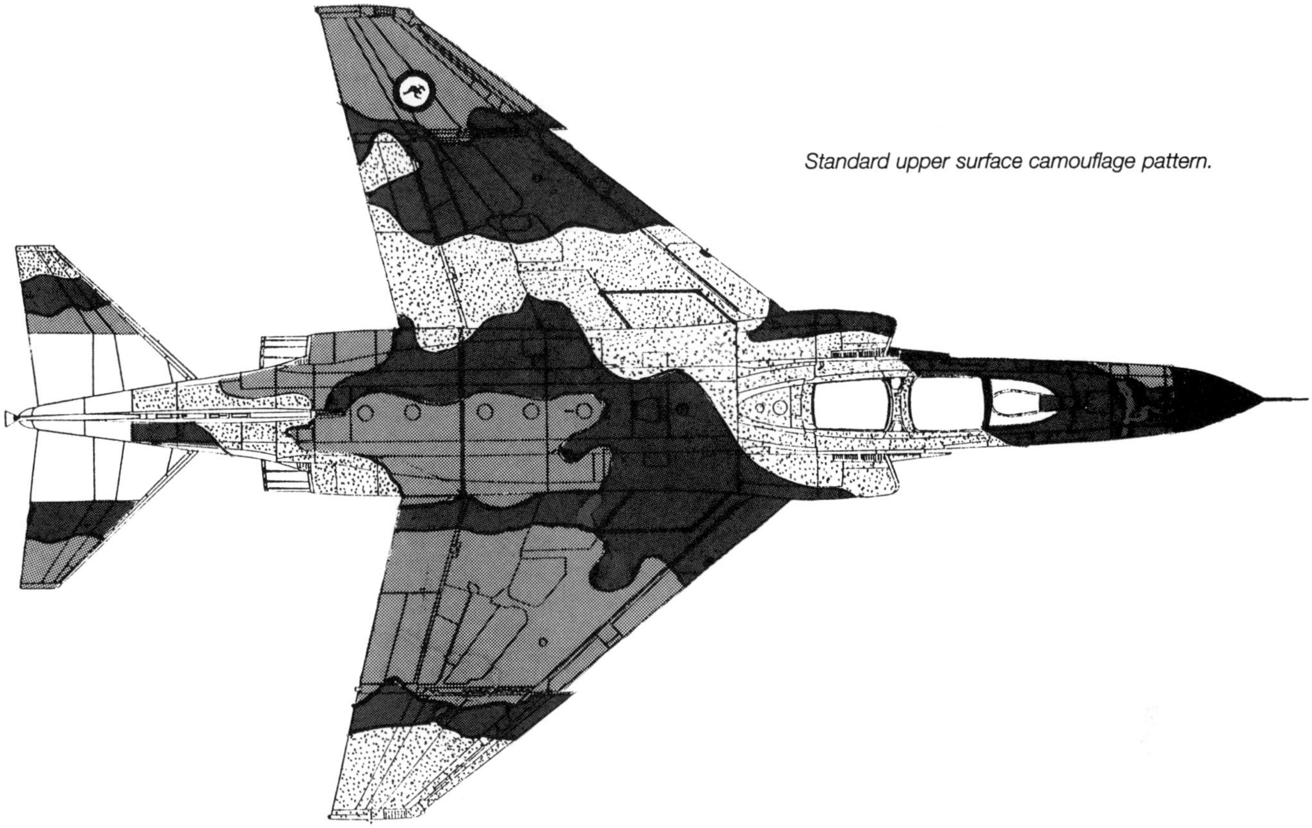

Standard upper surface camouflage pattern.

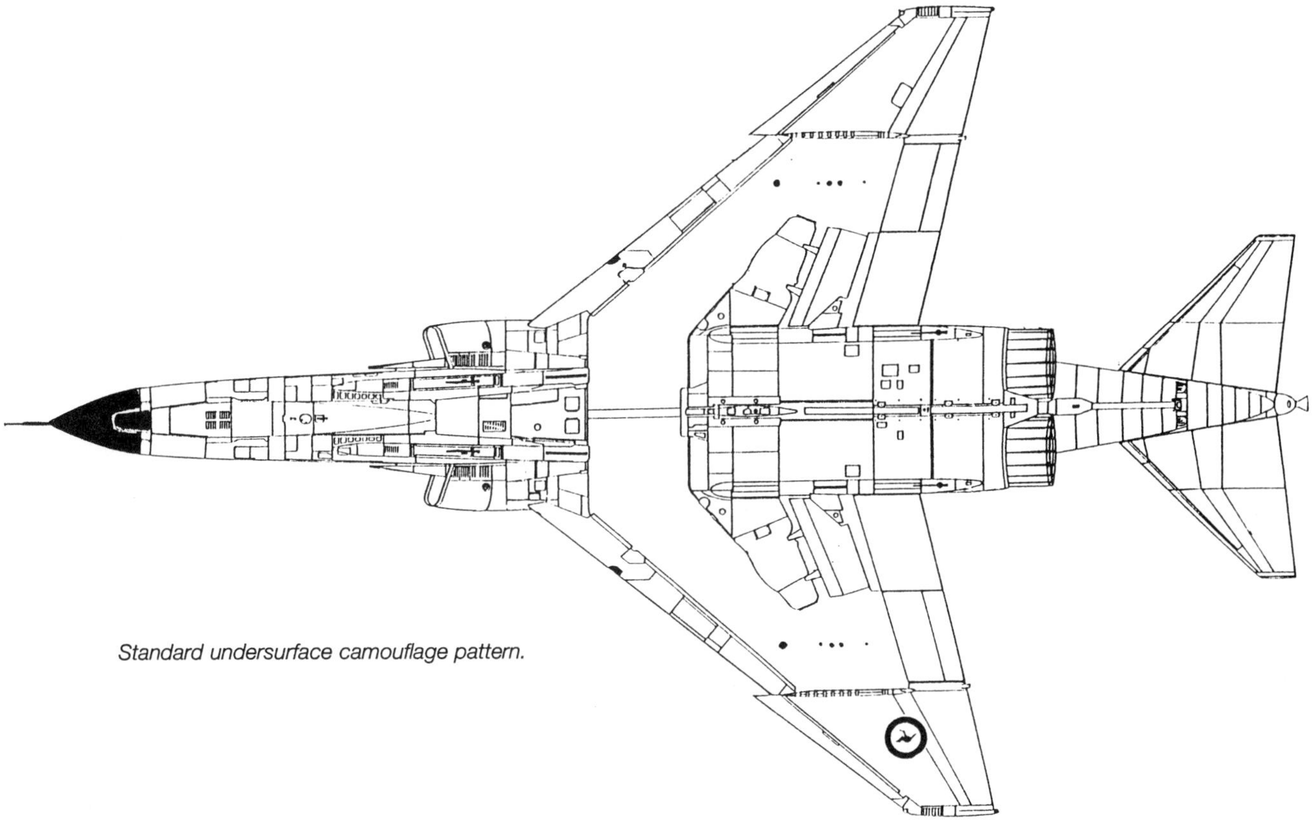

Standard undersurface camouflage pattern.

RAAF McDONNELL DOUGLAS F-4E PHANTOMS

Notes: Although the RAAF serial number prefix A69 was allocated to the Phantoms, this out of sequence number was used as a stores reference number only and reflected the USAF's own serial prefix. The aircraft retained their USAF serials during service with the RAAF. All were Block 43 aircraft (F-4E-43-MC).

An additional F-4E (USAF 67-0327) was presented to the RAAF Museum in 1989 and has been repainted in RAAF markings as 69-7234.

USAF No	Deliv	Disposal/Remarks
69-0304	09/70	returned to USAF 06/73
69-0305	09/70	returned to USAF 10/72
69-0306	09/70	returned to USAF 11/72
69-0307	09/70	returned to USAF 06/73
69-7201	09/70	returned to USAF 06/73
69-7202	09/70	returned to USAF 11/72
69-7203	09/70	crashed near Evans Head NSW 06/71
69-7204	09/70	returned to USAF 10/72
69-7205	09/70	returned to USAF 06/73
69-7206	09/70	nosewheel collapse on t/o 01/72, repaired by No 3 Aircraft Depot, reflown 09/72, returned to USAF 06/73
69-7207	09/70	returned to USAF 06/73
69-7208	10/70	returned to USAF 06/73
69-7209	09/70	returned to USAF 11/72
69-7210	10/70	returned to USAF 06/73
69-7211	10/70	returned to USAF 06/73
69-7212	09/70	returned to USAF 10/72
69-7213	09/70	returned to USAF 10/72
69-7214	10/70	returned to USAF 10/72
69-7215	09/70	returned to USAF 06/73
69-7216	09/70	returned to USAF 06/73
69-7217	09/70	returned to USAF 11/72
69-7219	09/70	returned to USAF 06/73
69-7220	10/70	returned to USAF 11/72
69-7234	10/70	badly damaged Amberley Qld 10/70, rebuilt by No 3 Aircraft Depot, reflown 09/71, returned to USAF 10/72

McDONNELL DOUGLAS
F/A-18 HORNET

A pair of F/A-18A Hornets (A21-8 and -18) from No 3 Squadron RAAF climb away. Each carries a pair of AIM-9L Sidewinder missiles on the wingtip stations. (DoD)

McDONNELL DOUGLAS F/A-18 HORNET

It seems that jet fighter/bombers which have been originally designed for operation from aircraft carriers and carry the brand name 'McDonnell Douglas' are just as happy operating from dry land and have found a worldwide market with land based air forces as well as the sponsoring naval service.

The McDonnell Douglas F-4 Phantom and A-4 Skyhawk are cases in point, although the US Navy's and Marine Corps' Skyhawk was never bought by the US Air Force, unlike the Phantom. McDonnell Douglas' most recent shipborne fighter, the F/A-18 Hornet, has also not been operated by the USAF but has found a steady market with several overseas air forces whose operations are entirely from *terra firma*.

The Royal Australian Air Force is one of these, the 75 Hornets it took delivery of between 1985 and 1990 replacing the Mirage as the service's front line fighter. Like most modern combat aircraft, the Hornet is expected to have a lengthy period of service with Australia of at least 20 years. During that time it will have provided Australia's primary air defence capability as well as a substantial short/medium range strike capacity and its importance to the RAAF and Australia is indicated by the lengthy period of evaluation which preceded its ordering.

For a relatively small country like Australia, the ordering of a new fighter is something which occurs only every two decades or so in these modern times. Such a fighter is usually the smaller nation's only such aircraft type, so the selection of the correct aircraft is of paramount importance as there is no backup, no 'second chance'.

The Hornet's virtues – its performance, its multirole versatility, its twin engined safety, its advanced systems and its weapons capability – made it the right aircraft for Australia and so far, five nations other than the USA: Spain, Canada, Kuwait, Switzerland and Finland.

Northrop Origins

Today's McDonnell Douglas F/A-18 began life in 1966 as a Northrop design study for medium weight multirole fighter combining high performance with low cost and offering an attractive follow on to the company's successful F-5/T-38 series of lightweight fighters and trainers. In between time there have been several changes of designation and/or name, a change in the aircraft's *modus operandi* (from strictly land based to carrier based and then to both) and indeed in the name of its manufacturer.

These early design studies included features which are now familiar on the Hornet – twin engines, modest wing sweep and the use of leading edge wing extensions (LEX) designed to help the wing maintain lift at high angles of attack. By 1968 the design incorporated twin vertical tail surfaces to maintain directional stability and later on automatic trailing and leading edge flaps were added to assist the LEX by effectively doubling the wing's lift coefficient when extended.

All these features are noted in the F/A-18 Hornet, but for the moment (from 1970) Northrop's design studies had resulted in a fighter called the P.530 Cobra, the name reflecting the 'hooded' appearance of the

The Northrop YF-17, unsuccessful candidate in the US Air Force's 'lightweight fighter' contest of the 1970s but then developed into the F/A-18 Hornet.

The YF-17 masquerading as a prototype F-18 Hornet, complete with 'NAVY' signwriting. McDonnell Douglas assumed leadership on the Hornet programme with Northrop acting as the main sub-contractor.

leading edge extensions which were similar to that exhibited by the reptile when angry.

Originally intended to be an international fighter in which ordering countries shared the development and manufacture, the Cobra found a more local potential customer in 1971 in the form of the US Air Force when it issued a Request For Proposals (RFP) for a new specialised and experimental day fighter combining the virtues of low weight and low cost. A modified P.530 design was submitted in early 1972 for this project, the USAF Lightweight Fighter (LWF) programme.

Four manufacturers submitted proposals of which two – General Dynamics and Northrop – were invited to build prototypes under the terms of what were called technology demonstration contracts, remembering that at this stage the intention of the LWF programme was to only explore new technologies and to provide a basis for future designs. The General Dynamics Model 401 entry became the YF-16 and Northrop's P.530 development the YF-17.

The two aircraft had fundamental differences despite being designed to the same specification. The most obvious one was the number of engines, a single Pratt & Whitney F100 turbofan engine of around 25,000lb afterburning thrust in the case of the YF-16 and two 15,000lb thrust with afterburner General Electric YJ101 turbofans in the YF-17. Both designs featured new technologies in construction, aerodynamics, avionics, weapons control and flight control systems. Both offered comparatively modest takeoff weights in their initial forms; 17,500lb (7,938kg) in the case of the YF-16 and about 21,000lb (9,525kg) for the YF-17.

The two YF-16 prototypes were the first in the air, the aircraft flying in February and May 1974. The two YF-17s followed on 9 June and 21 August of the same year, but before that year was out, the LWF programme underwent a fundamental change.

At that point it was decided to change the contest into one for a new air combat fighter (ACF) to complement the McDonnell Douglas F-15 Eagle in the USAF. Evaluations were completed by January 1975 and in that month the YF-16 was selected as the winner, the reasons given being superior performance, engine commonality with the F-15 and lower cost, although political considerations were also probably involved as General Dynamics' Fort Worth facility was in need of work at the time.

The production F-16 Fighting Falcon has nevertheless proved to be a fine aircraft and has subsequently been built in large numbers not just for the USAF but also for more than a dozen other nations.

Despite the loss of the lucrative ACF programme, the YF-17 had demonstrated some remarkable characteristics during its 345 flight test hours to that point. It had reached an angle of attack (AOA) of 63 degrees at an indicated airspeed of just 20 knots (37km/h) and an AOA of 40 degrees with a sideslip angle of 38 degrees, attitudes which were beyond the F-16's capability. In addition, the YJ101 engines had shown no unwelcome characteristics and the larger design of the YF-17 meant it had greater potential than its competitor in the attack role as well as the greater security offered by two engines.

Enter the Navy

Even though Northrop had not won the competition against the F-16 for the USAF fighter order, all was not lost. In early 1974, the US Navy had defined its requirement for a new fighter/bomber to replace the Vought A-7E Corsair II and McDonnell Douglas F-4J Phantom II. Responses to this requirement were received from six manufacturers but in August 1974 the US Congress ordered that the Navy's new fighter should be a derivative of either the YF-16 or YF-17.

The new programme was called NACF (Naval Air Combat Fighter), and as the US Navy prefers dealing with manufacturers experienced in naval aviation – and indeed insisted that the prime contractor on the programme be such a manufacturer – General Dynamics joined forces with Vought and Northrop with McDonnell Douglas.

The 'navalisation' of the P.530 design brought with it a new company designation – P.630 – as well as the introduction of a McDonnell Douglas model number (267) reflecting that manufacturer's increasing influence on the aircraft's design.

In May 1975 the Northrop/McDonnell Douglas proposal was selected for further development over the GD/Vought effort based on the YF-16, the decision vindicating Northrop's faith in a design concept which had been born nearly a decade earlier. There was another side to the coin for

Two views of the prototype Hornet, first flown on 18 November 1978. The word 'MARINES' was painted on the starboard side of the fuselage and 'NAVY' on the port side. (Northrop/MDC)

Interesting angles on a US Navy Hornet as it sits on the flight deck of its carrier. Note that only the starboard main flap is extended. (MDC)

The second Hornet, which first flew on 12 March 1979. All 11 examples of the development batch were flying by March 1980. (MDC)

Northrop though, and that was the fact that McDonnell Douglas would be the prime airframe contractor on the project (including responsibility for final assembly) and that Northrop would be a mere sub contractor, albeit the major one with responsibility for 30 per cent of the airframe development work and 40 per cent of airframe manufacture.

Broken down, this left Northrop with the manufacture of the rear and centre fuselage as well as responsibility for most of the systems while McDonnell Douglas would manage the programme, manufacture the remainder of the airframe, the undercarriage and arrestor equipment and look after the cockpit, avionics and flight controls. The aircraft would be assembled at McDonnell Douglas' St Louis plant.

The NACF programme initially called for about 800 aircraft but this was later increased to 11 development and 1,366 production aircraft to equip 24 US Navy attack and six fighter squadrons plus 12 US Marine Corps fighter squadrons with 332 aircraft for reserve units. This total was subsequently reduced to 1,168 aircraft.

There was some controversy surrounding the selection of the P.630/Model 267, as some understood that the US Congress' order that the winning aircraft should be selected from the ACF contestants implied the winner of that contest should be the one developed for the Navy.

What would become the Hornet was certainly superior to the opposition design, meeting all but one of Navy's requirements, the critical one of approach speed. This would be sorted out during flight testing, but in the meantime General Dynamics and Vought lobbied hard to have their aircraft reconsidered.

The third Hornet overflies the carrier USS America *before making the type's first sea landing in November 1979. (MDC)*

The first of a pair of two seater Hornets (at this stage designated TF-18s) which were included in the development batch. (MDC)

Matters of Designation

It might be useful at this point to sort out some of the designations which have been applied to the Hornet over the years. The original intention was to have separate fighter and attack variants so the designations F-18 and A-18, respectively, were applied to them. In addition, the two seat operational trainer variant was to be called the TF-18.

The combination of these roles within a single airframe was subsequently found to be possible by the use of different external loads and thanks to the versatility of the radar selected for the aircraft, the pulse doppler, multi mode Hughes APG-65.

The result was the new designations F/A-18A for the initial single seat model and F/A-18B for the two seater.

These designations initially applied only to the naval variants of the Hornet, while the generic designation F-18L was used for land based versions promoted by Northrop. One of the YF-17 prototypes was demonstrated to Canada in 1978 (before the first F-18 had flown) and dubbed F-18L in Northrop's promotional material but in the event Canada's (and other customers') Hornets were essentially unchanged naval variants using the normal McDonnell Douglas designations.

The issue of which of the partners would market and apply their name to which Hornets wasn't finally resolved until 1985 following litigation, the resultant agreement confirming McDonnell Douglas as prime contractor and Northrop as principal sub contractor on all existing and future versions of the aircraft. Before this was sorted out it was possible for the two manufacturers to compete against each other for an order with different versions of the same aircraft!

The result of this has been that there are no separate naval and land based Hornet variants. All are basically the same aircraft with folding wings and arrestor hooks.

From Cobra to Hornet

Although sharing the same overall configuration as the Cobra/YF-17, there was considerable redesign involved in turning the design into the definitive F/A-18 Hornet. Although its overall dimensions were only slightly greater than the YF-17's, the Hornet featured an enlarged fuselage to accommodate more fuel and the APG-65 radar, the wing area was increased by 15 per cent to

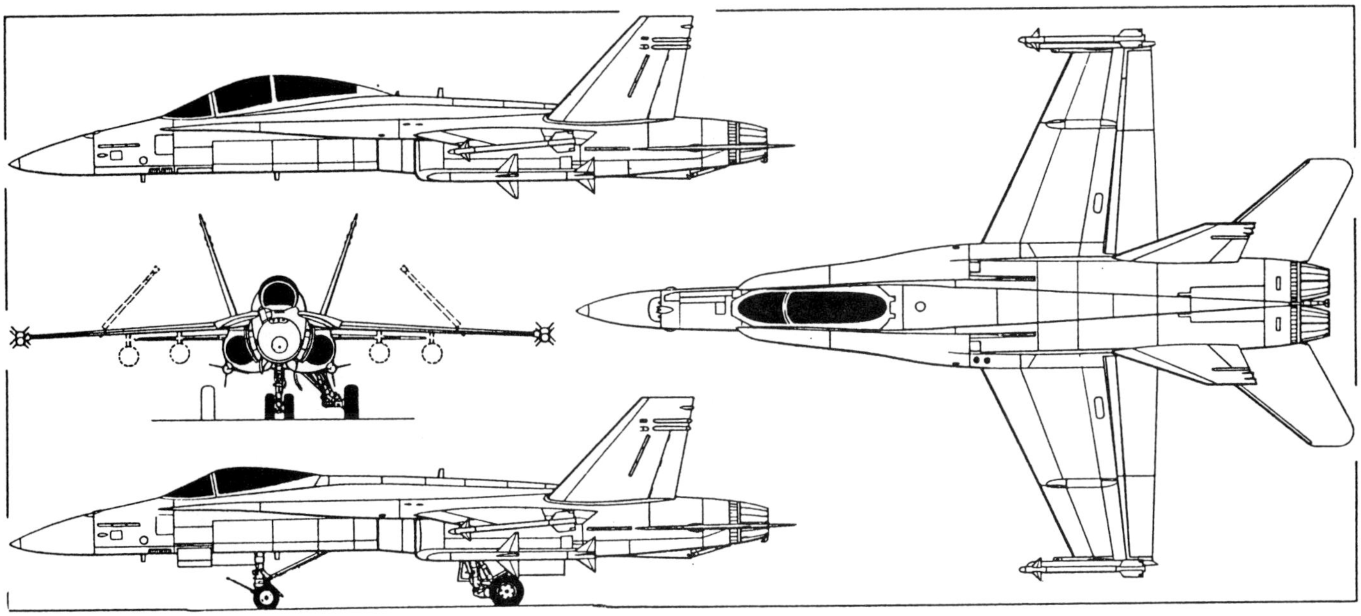

maintain wing loading in the face of weight increases (and made to carry fuel) and the airframe structure strengthened to absorb the loads imposed by catapult launches and carrier landings as well as the substantial increases in weight.

The Hornet's maximum permissible weight of 56,000lb (25,400kg) is some two-and-a-half times greater than the YF-17's and this factor – along with the need to cope with rates of descent of up to 24ft/sec (7.32m/sec) resulted in a much stronger undercarriage.

The General Electric YJ-101 engine was fitted with a larger fan, turning into the F404 which offered about 16,000lb thrust with afterburner, provision for the US Navy's probe and drogue flight refuelling system was made, complete with retractable probe and the necessary wing folding mechanism, and catapult launching attachments and arrestor hook were incorporated.

The armament provision of the YF-17 remained intact (a fixed 20mm M61 multi barrel cannon plus stores stations) although the latter were reorganised with the deletion of two of the original six underwing stations and the introduction of two new stations on the lower fuselage along with wingtip and centreline stations, making nine in all.

The salient features of the Hornet are recorded in the accompanying data table, but it is worth noting some of the advanced features incorporated into its design, particularly the radar/fire control system, the cathode ray tube (CRT) cockpit displays, the fly-by-wire controls, head up display (HUD), hands on throttle and stick (HOTAS) controls, self assessing maintenance software and the use of advanced materials in the airframe's construction.

By weight, the Hornet's structure contains 49 per cent aluminium alloy, 16.7 per cent steel (undercarriage, tailplane spars and ribs, some wing ribs), 12.9 per cent titanium (vertical tail attachments, wing fold fittings, wing root skins, some vertical tail skinning), 9.9 per cent graphite/epoxy (tail surface skins, wing torque box, flaps, dorsal spine) and 10.9 per cent 'others' (canopy glass, tyres, radome etc).

The two seater Hornet retains full operational capability, the rear cockpit losing only the HUD. To accommodate the extra crew member, about 6 per cent of the internal fuel capacity is lost with the removal of the forward fuselage cell.

Testing Time

The development batch of 11 Hornets comprised nine single seaters and a pair of trainers (numbers 7 and 10 and still referred to as TF-18s at this stage), the first of which (serial number 160775) was rolled out of the St Louis factory on 13 September 1978 and flown for the first time on 18 November. It carried the legend 'MARINES' on one side of the fuselage and 'NAVY' on the other.

The remainder of the development batch followed over the next 16 months and all 11 were flying by March 1980. The first sea trials were carried out in November 1979 aboard the USS *America* and the first US Navy units to receive Hornets in 1980 were VX-4 fighter test and evaluation squadron and VX-5 attack and evaluation squadron. VFA-125 (the Hornet Fleet Readiness Squadron) received its first aircraft in February 1981.

Testing was not an entirely smooth process with several problems needing to be sorted out. Weight was up, performance was down, combat radius was less than hoped for, the wing was flexing in the roll thus reducing its rate, engines were failing and there were fatigue problems with the vertical tail attachment structure. The air-

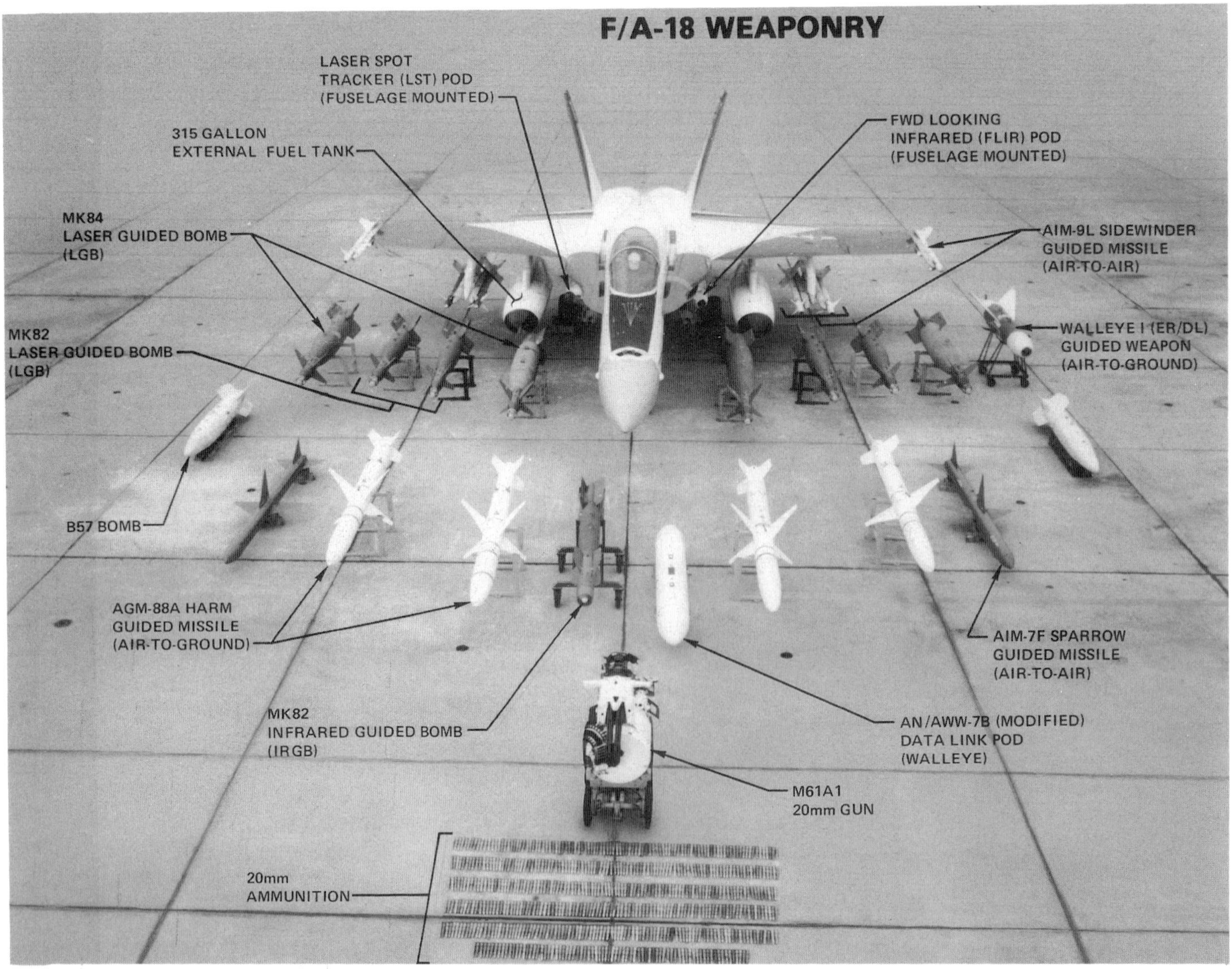

Nose detail of an F/A-18B showing the inboard portions of the leading edge flaps, engine intakes and gun muzzle aperture with the gun gas venting intakes alongside.

craft received some bad press through all this but most problems were solved.

A drag reduction programme restored performance to acceptable levels, the wing was stiffened to prevent flexing, aileron size was increased by a substantial 36 per cent (and resulted in lower approach speeds as they are drooped on the approach) and the differential tailplane authority was increased, and a fix was found for the rear fuselage fatigue cracking although a better solution was installed later when a LEX fence was fitted which modified the vortex pattern and reduced stresses in the rear fuselage/tail attachment areas.

The F404 engines suffered compressor blade fatigue failures and resulting titanium fires, this problem being solved by the fitting of modified blades and a fire retardant mat around the casing.

Another problem was that of takeoff rotation speed which was found to be considerably higher than predicted. The cause was traced to the fact that the Hornet's mainwheels had been moved backwards compared to the YF-17's so as to remove any possibility of the aircraft tipping on its tail whilst being towed backwards on a dipping and swaying flight deck. This problem was solved by increasing the area of the tailplane, reducing the leading edge flap setting for takeoff from 30 degrees to 15 degrees and automatically toeing in the rudders when there is weight on the wheels. The result was a startling 30-35 knots (56-65km/h) reduction in rotation speed.

Production and Variants

The first operational Hornet unit was Marine Corps Squadron VMFA-314, which received its first aircraft in July 1982 and after working up was declared operational in January 1983. The first Navy operational unit was VFA-113 from October 1983. This squadron, along with VFA-25, began the Hornet's first operational deployment aboard the USS *Constellation* in February 1985.

Production of the F/A-18A (380 built including prototypes) and the F/A-18B (41 built) for the USN/USMC ended in 1987 and was replaced by the F/A-18C and D, single and two seat variants, respectively.

Improvements incorporated into these models include provision for up to six AIM-120 AMRAAMs (advanced medium range air-to-air missiles), up to four imaging infra-red Maverick missiles, jamming equipment and improved avionics and computer.

A developed subvariant, the F/A-18C/D Night Attack was first flown in 1988 and delivered from late 1989. This version features pilot's night vision goggles, a thermal imaging navigation set, external targeting forward looking infra-red (FLIR) pod and colour multi function cockpit and moving map displays. The Marines' F/A-18D has no control column in the rear cockpit but two sidestick weapons controllers while the Navy's Ds can be converted to full dual control for training. The F/A-18D (RC) is a simple reconnaissance development with a twin sensor package replacing the gun in the nose.

McDONNELL DOUGLAS F/A-18A HORNET

Powerplants: *Two General Electric F404-GE-400 low bypass ratio turbofans each rated at 10,600lb thrust dry and 16,000lb thrust with afterburner; engine dry weight 2,180lb (989kg). Internal fuel capacity approx 1,600US gal (6,057 l) in self sealing tanks, provision for one to three 330US gal (1,249 l) external tanks, maximum combined capacity approx 2,590US gal (9,804 l). Retractable flight refuelling probe in upper starboard nose.*

Accommodation: *Pilot only (F/A-18A) or two pilots (F/A-18B) on Martin Baker SJU-5/6 zero-zero/600 knot ejection seat.*

Airframe: *Semi monocoque fuselage primarily of light alloy with graphite/epoxy access doors; titanium firewall between engines; pressurised cockpit section; airbrake on upper fuselage surface between fins; fixed inlet ramps.*

Multi spar wing of mainly light alloy and graphite/epoxy construction; boundary layer control by wing root slots; full span leading edge manoeuvring flaps extend to 30deg; single slotted trailing edge flaps extend to 45deg; ailerons droop to 45deg for low approach speed; leading and trailing edge flaps computer controlled for optimum lift or drag; wing root leading edge extensions for high angle of attack capability; wings fold inboard of ailerons.

Tail surfaces of mainly graphite/epoxy construction. All moving horizontal tailplanes (stabilators); twin canted fins and rudders; tail surfaces operate collectively and differentially.

Retractable tricycle undercarriage stressed for carrier operations. Twin wheels on forward retracting nosegear; single wheels on main undercarriage which retracts rearwards, rotating through 90deg to stow horizontally inside lower surface of engine air ducts. Tyre pressures (land operations) mainwheels 200lb/sq in (13.79 bars), nosewheels 150lb/sq in (10.34 bars). Arrestor hook under rear fuselage, nosewheel catapult attachment (US Navy).

Systems: *Two independent hydraulic systems of 3,000psi; quadruplex digital fly-by-wire control system with direct electrical backup to all surfaces and direct mechanical backup to stabilators; air conditioning, electrical, oxygen and fire detection/extinguishing systems; Garrett auxiliary power unit (APU) for aircraft starting.*

Avionics: *Hughes AN/APG-65 multi mode digital air-to-air and air-to-ground tracking radar capable of tracking ten targets simultaneously and displaying eight of these to pilot. General Electric quadruple redundant flight control system; two AYK-14 digital computers; Litton AN/ASN-130A inertial navigation system; two Kaiser multi functions CRTs (cathode ray tubes); central Ferranti/Bendix CRT display and HUD (head up display); digital data recorder for Bendix maintenance recording system; dual UHF/VHF communications system; VOR/ILS (ACLS – automatic carrier landing system on USN/USMC aircraft); TACAN (tactical air navigation system); HF radio; data link; integral ECM (electronic countermeasures); automatic moving map; ALR-67 radar warning receiver.*

The APG-65 radar allows rapid transition between surface attack and air combat missions. There are three pilot selectable Air Combat Manoeuvring (ACM) modes in which the radar automatically locks on and tracks airborne targets – Wide Acquisition, Vertical Acquisition and Boresight. A fourth ACM is initialised when the pilot selects the internal gun and a fifth mode – Auto Acquisition – is also available.

Air-to-ground modes include a beam ground map mode with two additional levels of 'zoom' capability providing outstanding cockpit imagery. Additional features, linked with weapons/fire control system are air-to-ground ranging, fixed target track and precision velocity update, sea surface targeting and ground moving target identification/track.

Dimensions: *Wing span 37ft 6in (11.43m); span over missile 40ft 4.75in (12.31m); wing chord at root 13ft 3in (4.04m), at tip 5ft 6in (1.68m); wing aspect ratio 3.5:1; width wings folded 27ft 6in (8.38m); length 56ft 0in (17.07m); height 15ft 3.5in (4.66m); tailplane span 21ft 7.25in (6.58m); wheel track 10ft 2.5in (3.11m); wheelbase 17ft 9.5in (5.42m).*

Weights: *Max t-o 56,000lb (25,400kg); typical t-o weight fighter mission 36,710lb (16,652kg), attack mission 49,224lb (22,328kg); empty 23,050lb (10,455kg).*

Armament: *One 20mm M61 six barrel cannon in nose with 570 rounds; nine external weapons station with combined capacity of 17,000lb (7,711kg). Two wingtip stations for AIM-9 Sidewinder air-to-air missiles; two outboard wing stations for AIM-9 Sidewinder or AIM-7 Sparrow air-to-air missiles or air-to-ground ordnance (AGM-84 Harpoon, AGM-65 Maverick, laser guided bombs, general purpose bombs, cluster bombs etc); two inboard wing stations for external fuel tanks or air-to-ground weapons; two nacelle fuselage stations for AIM-7 Sparrow missiles or FLIR (forward looking infra-red) pods; one underfuselage centreline station for external fuel or weapons; chaff/flare dispenser.*

Performance: *Max speed more than 1,190mph (1,910km/h) or Mach 1.8 at 37,000ft; max speed (intermediate power) more than 660mph (1,062km/h) or Mach 1 above 30,000ft; initial climb rate (50% fuel, two wingtip AAMs) 60,000ft (18,288m)/min; combat ceiling approx 50,000ft (15,240m); takeoff run less than 1,400ft (427m); approach speed 154mph (248km/h); combat radius of action (fighter mission) more than 460 miles (740km), (attack mission) 662 miles (1,065km); unrefuelled ferry range more than 2,300 miles (3,700km).*

Hornets in service with the US Navy about to be launched (top) and at the point of launch (below) from the USS Constellation. *The lower photograph well illustrates the Hornet's large flap area when the ailerons are drooped.*

The first F/A-18C taking off from St Louis on 3 September 1987. This improved Hornet features provision to carry advanced medium range air-to-air missiles (AMRAAM), Maverick missiles, jamming equipment and updated avionics. (MDC)

The US Navy's aerobatic team, the Blue Angles, swapped its Skyhawks for Hornets when the latter became available (top); the first Hornet export customer was Canada (bottom) which ordered 138 aircraft in 1980. (Paul Merritt)

From early 1992 Hornets have been delivered with F404-GE-402 engines offering about 10 per cent more thrust than the original -400s, while improved Hughes AN/APG-73 radar is scheduled to be fitted to production aircraft from 1994. The first APG-73 equipped Hornet flew in April 1992.

The 1,000th Hornet (an F/A-18D for the US Navy) was handed over in April 1991, by which time the fleet had amassed well over a million flying hours. By mid 1992 total Hornet production had exceeded 1,200 aircraft of which 900 were for the USN/USMC (including prototypes) and 325 were for export.

Future Hornet versions being planned at the time of writing include the F/A-18E and F with longer fuselage, increased weights, larger wing and tail surfaces, revised inlets, increased fuel, more powerful engines and two extra stores stations. This Hornet is due to fly in 1995 and enter service in 1999.

Exports

The first export customer for the Hornet was Canada which ordered 114 CF-18As and 24 CF-18Bs in April 1980. The numbers were later revised to 90 and 48 respectively, still leaving a total of 138 and deliveries were made between October 1982 and September 1988.

Canadian Hornets differ from the US versions in several minor ways – they have an ILS (instrument landing system) in place of naval aircraft's ACLS (Automatic Carrier Landing System), a spotlight on the port side of the fuselage for night time visual identification of targets, provision for rocket pods and a cold weather land survival kit for the pilot.

Australia was the Hornet's second export customer (as discussed in detail in the following chapters) while the third was Spain, which ordered 60 EF-18As and 12 EF-18Bs in May 1983. Deliveries began in 1986 and ended in 1990. Spanish Hornets have the ACLS and catapult launch bar deleted and ILS added.

Other export customers to date are Kuwait, which ordered 32 Cs and eight Ds in 1988 with deliveries beginning in early 1992; Switzerland is due to take delivery of 26 Cs and eight Ds from 1995 and Finland will start receiving seven Ds (built by McDonnell Douglas) and 57 Cs (assembled by Valmet from kits) in the same year. South Korea was a 'lost' customer, ordering 120 F/A-18C and Ds in late 1989 but cancelling the contract just over a year later in favour of the Hornet's old rival, the General Dynamics F-16.

The F/A-18 Hornet has had an interesting, twisting and turning development history which has resulted in a mature and effective warplane which should remain in production in one form or another for some time yet.

The aircraft has seen action in Libya and the Gulf War, the latter conflict involving aircraft from the American carriers *Midway*, *Saratoga*, *Independence*, *America*, *Dwight D Eisenhower* and *Theodore Roosevelt* flying both offensive and defensive missions. Canadian Hornets were also involved in the conflict, CF-18s from No 409 Squadron (normally based in Germany) operating from Qatar in defence of the Canadian warships also helping the Allied cause. The round of strikes against Iraqi targets in early 1993 also involved US Navy Hornets.

RAAF F/A-18B Hornet A21-102 taxies in at RAAF Fairbairn, Canberra. (DoD)

THE TFF PROJECT

Australia's Tactical Fighter Force (TFF) Project – the selection and production of a replacement for the Royal Australian Air Force's fleet of Mirage III fighters – has undoubtedly been one of the most expensive and important government programmes of the latter part of the 20th century.

The selection of a new fighter for a relatively small air force such as Australia's is an extremely significant event as it is something which these days occurs infrequently, the winning aircraft expected to remain in service for at least two decades and in the meantime be capable of performing its role(s) effectively for its full service life.

Apart from properly fulfilling the operational roles intended for it, there are other factors which must be taken into account before a particular type is selected. A vitally important one is that of local industry participation, Australia having a long history of being involved in at least the assembly and part manufacture of several of its major combat aircraft since World War II – Vampire, Sabre, Mirage, Lincoln and Canberra among them.

Local industry participation is not limited to just the airframe but also the powerplant and numerous smaller but no less important items of equipment. In the case of the TFF Project winner – the McDonnell Douglas F/A-18 Hornet – some 17 Australian companies were and are involved, performing services ranging from the supply of hydraulic pumps and electronic components to the manufacture of metal and advanced composite materials and the final assembly of the aircraft.

Further, involvement in an advanced 'state-of-the-art' aircraft such as the Hornet has significant longer term industrial benefits for the participating nation due to 'technology transfer' and the associated spin-offs.

The Search Begins

More than a decade passed between the first investigations into a Mirage replacement for the RAAF and the placing of orders for the Hornet in October 1981, although it should be pointed out that early investigations were interrupted by changes of government and policy and other issues. The definitive selection process began only in 1977 and two years later the list of possible candidates had been reduced to two.

There was some discussion and investigation as early as 1970 and 1971 when the RAAF first expressed interest in the forthcoming McDonnell Douglas F-15 Eagle for service in the latter part of the 1970s, while the Mirage F1 was also mooted at around the same time with talk of licence production in Australia. At that time it was considered that the RAAF's Mirage IIIs would need to be replaced from 1980.

In December 1971 the RAAF issued an Air Staff Requirement for a new fighter with responses from manufacturers required by the end of February 1972. The response was overwhelming, to such an extent that the deadline for manufacturers to submit proposals was extended by a month.

In April 1972 the list of aircraft being studied included just about every current or proposed fighter in the western world: Saab Viggen, McDonnell Douglas F-15 Eagle, Northrop P.530 Cobra, Lockheed CL-200 Lancer (an advanced F-104 Starfighter derivative), Dassault Mirage F1, Grumman F-14 Tomcat and Panavia MRCA Tornado.

Ever innovative when it comes to marketing its aircraft, Dassault attempted an interesting approach by offering Australia a co-production deal on the Mirage F1 whereby Australian industry would manufacture certain components for the world market with absolutely no obligation for the RAAF to order the type.

It looked as if the proposal was going to be accepted by the Australian government for a time in late 1972 despite the RAAF's misgivings about how it would more or less oblige the service to take F1s whether it wanted them or not. Of course this prospect had never entered the Dassault marketing team's thoughts! Discussion on the subject ended shortly afterwards when the Department of Defence expressed its own

An early proposal made to Australia in the search for a Mirage III replacement came from Dassault, which suggested that Mirage F1 parts be manufactured in Australia for the world market regardless of whether or not the RAAF bought the type. The F1 remained in contention for some years. (AMD)

misgivings, saying the proposal did not take the RAAF's needs into full consideration.

In March 1973 AVM F S Robey and team went overseas to study the Mirage F1, the Eagle, the Cobra and the Viggen but recommended it might be wise to wait until other types became available. Things ground to a halt the following August when it was announced by the recently installed Labour government that the project was to be deferred. Instead, one Mirage squadron would be disbanded and the type's life extended for service into the 1980s.

New Influences

The US Air Force's lightweight fighter competition of the mid 1970s (as discussed in the previous chapter) introduced new factors into the on again-off again search for a new RAAF front line fighter. This competition pitted the General Dynamics F-16 and the Northrop Cobra/YF-17 against each other, the F-16 eventually winning and the YF-17 being developed into the F-18 Hornet for the US Navy.

The F-16 was also selected for a multi national fighter replacement programme in several European air forces, the deal including co-production arrangements. France had meanwhile developed an all new Mirage, the delta wing 2000, which would be ordered into production for the *Arm'ee de l'Air*.

By 1975 there was new activity within the government on the Mirage replacement question, activity which saw the true start of the programme which eventually resulted in the F/A-18 Hornet programme. The defence minister indicated that the type which would fill the RAAF's tactical fighter force requirement might be announced the following year and would cost $500m.

Another change of government in late 1975 brought with it increased awareness of defence issues, resulting in some concrete moves being made on the fighter issue including the issue of a Request for Proposal, resulting in responses from 11 manufacturers. Throughout 1976 it was anyone's guess which aircraft was on the top of the list. Speculation was rife, newspapers publishing various guesses as to the short list during the course of the year, even though evaluation proper didn't begin until 1977.

The F-16, Viggen, Mirage F1, F-15, Tornado and even the Sepecat Jaguar (primarily a ground attack aircraft), were all mentioned in the same breath early in 1977, while by mid year one newspaper was claiming a short list of three aircraft: F-14, Viggen and Tornado; in late 1977 another newspaper reported the short list as being F-15, F-16, Jaguar and – of all things – the Fairchild A-10 dedicated ground attack aircraft. Not to be outdone, yet another newspaper report of the same period named the F-15, F-16, Mirage 2000 and for the first time, the F-18 Hornet.

Contenders for the Australian TFF contract included (top to bottom) the Northrop YF-17 (seen here in the guise of the proposed F-18L), the General Dynamics F-16 Fighting Falcon and the Saab Viggen.

A Concrete Programme

In the meantime, some definite and non-speculative progress had been made, Prime Minister Malcolm Fraser indicating that it should be an American aircraft which replaced the Mirage. Also in 1977, a Defence Aircraft Industry Steering Group DAISG) was established to "consider the ways in which the resources of the local aircraft industry should be utilised and developed to contribute most effectively to supporting the Tactical Fighter Force throughout its life", and to "consider the various alternatives available in negotiations with potential suppliers to maximise the value of local industry participation and examine the influence of phasing on the negotiating position".

With this important move, the acquisition of a new fighter for the RAAF was now formally a matter of 'when' rather than 'if'. Which aircraft would be purchased was still a matter of some conjecture.

By April 1978 the list didn't appear to be getting any shorter, Defence Minister Jim Killen advising parliament that the F-15, F-16, McDonnell Douglas F-18A *and* Northrop F-18L (refer previous chapter), Tornado and Mirage 2000 were all being examined.

In May 1978 a Department of Defence team went overseas to study the aircraft of this rather lengthy shortlist, and the government talked about an initial purchase of just 25 to 30 aircraft at a cost of $500m, a quantity seemingly at odds with the ideals of Australian industry participation and the all important industrial offsets which would need to be part of the deal.

The Department of Defence began recruiting companies interested in participating in the New Tactical Fighter (NTF) programme as it was now referred to in November 1978, while in 1979 evaluation teams went to the USA and France to study operational capabilities, engineering aspects of the competing aircraft and to study Australian industry participation.

These studies resulted in the Mirage 2000 and F-18L being dropped from the equation, leaving, in November 1979 a genuinely 'short' shortlist with just two types on it – the F-16 and F-18. But neither aircraft – said the defence minister – fully met RAAF requirements.

The Decision is Made

Evaluation of the relative merits of the final pair of contenders continued into 1980 but there were some more delays to endure yet, the first deferring the decision until after October's general election (the government was returned) and another, in December, which referred to 'substantial technical problems and shortcomings'. These included the F-16's missiles and certain structural and aerodynamic issues with the Hornet.

Finally, on 20 October 1981, Defence Minister Killen announced that the McDonnell Douglas F/A-18 Hornet would replace the Mirage III in RAAF service. The order was for 75 aircraft of which 57 would be

Three more TFF contenders (top to bottom): Dassault Mirage 2000, Panavia Tornado and Grumman F-14 Tomcat. Other more fanciful suggestions included the SEPECAT Jaguar, primarily ground attack aircraft and the Fairchild A-10 Thunderbolt II, very definitely a ground attack/close support type.

A standard marketing tool is to present illustrations of aircraft in the potential customer's colours, witness this rendering of a pair of 'RAAF' F-15 Eagles dated April 1976. (MDC)

single seaters and the remaining 18 operational trainers. In addition, 168 General Electric F404 engines were ordered.

Mr Killen commented on the importance of the programme: "The Tactical Fighter Force Project is a major national enterprise. It will involve not only the RAAF, and my department, but a host of contractors and sub contractors in Australia's defence related industry".

The major Australian industry participants were nominated as the Government Aircraft Factory (final assembly and flight testing) and the Commonwealth Aircraft Corporation (engine assembly and testing). Both also manufactured airframe components and were supported by numerous other Australian companies. The manufacturing programme is discussed in the following chapter.

And the cost of the Hornet programme? It was estimated at $2.43bn when the order was announced but was steadily upwardly revised after that, mainly due to the constantly falling value of the Australian dollar against the US dollar (some 40 per cent) in the following years. In 1983 the figure was put at $3.4bn then $4.4bn in 1985 and $5.5bn in 1988, including that which had to be spent on new facilities and the upgrading of existing ones and establishing the infrastructure necessary for the introduction of the most advanced aircraft yet to enter service with the RAAF.

Programme Management

The overall management of the programme was handled through the newly established Tactical Fighter Project Office (TFPO) within the Department of Defence, its staff of 100 mainly working at various postings in the USA – in Washington, at McDonnell Douglas' St Louis plant and at various US Navy air and weapons test facilities. For its part, the US Navy opened its first overseas Naval Plant Representative Office (NAVPRO) in Melbourne in 1982, the office acting as a kind of 'watchdog' over the programme. Both the NAVPRO and the local McDonnell Douglas office were paid for by the Australian government.

The NAVPRO office was commanded by a US Navy officer (Commander Mike Kearney at first then Commanders Koelher, Dyer, Freedman and Henry after him) and staffed by both American and Australian military and civilian personnel. Its main roles were to formally accept the completed aircraft from McDonnell Douglas and engines from General Electric; to test fly the aircraft on behalf of the customer (the US Navy – see below); to monitor the quality of the Australian product; to audit Australian industry bids; to monitor any factors (industrial unrest and failing to meet schedules, for example) which may affect the final product; and to assist Australian industry in meeting the requirements of US defence business, acquisition regulations and so on.

Australia's Hornets were purchased under a US Foreign Military Sales programme rather than as a direct purchase from McDonnell Douglas. This resulted in a necessarily complicated arrangement whereby the Hornets, their engines and associated parts and equipment which would eventually fly in RAAF colours were ordered by the US government through the US Navy from McDonnell Douglas and General Electric in the case of the engines. The aircraft, when built, were delivered to the US Navy.

McDonnell Douglas and GE, in turn, sub contracted work to companies like Northrop, which was responsible for the aft fuselage section and Hughes, which made the APG-65 radar. Northrop and Hughes had sub contractual agreements with local companies (CAC, GAF, Philips etc) who also had sub contracts with McDonnell Douglas and GE.

Local industry had contracts with the Australian government, thus completing the circle. As for the completed Hornets, they remained the property of the US government for only a few minutes after pre delivery test flying (usually involving about five flights) was completed. They were then handed over to the RAAF.

One of the first F/A-18Bs over Malaysia not long after being introduced into service.

General Dynamics prepared this artist's impression of an F-16 carrying the markings of No 76 Squadron RAAF and overflying Canberra.

The winner! The McDonnell Douglas F/A-18 Hornet, selected for the RAAF in October 1981.

Involving Australian Industry

The aims of the Defence Aircraft Industry Steering Group (DAISG) were mentioned earlier in this chapter. Its members included relevant senior public servants and two senior air force officers, AVM John Cornish (Chief of Air Force Materiel) and AVM Lyn Compton (Chief of Air Force Technical Services).

The DAISG reviewed six wide ranging options for Australian industrial participation in the TFF programme, ranging from the extremes of nil industry involvement to developing a new indigenous design. For each option, the level of industrial support which would flow from the programme, the costs, the workload, the risks and the schedule were assessed.

The group concluded: "An industry programme which included final assembly and test of the aircraft, manufacture of a representative range of airframe sub-assemblies, engine components and equipments for co-production and export would enhance the industry's capability, support the Defence Force and make possible a degree of self reliance representing a reasonable balance between costs and benefits.

"There would be advantages in basing plans for Australian industry involvement in the programme on the total number of aircraft likely to be ultimately acquired, and, the programme could add six months on to the lead time for acquiring aircraft".

These and other conclusions (published in 1978 when the Mirage 2000, F-18A and L and F-16 were in the running) formed the basis of the development of the industry programme, which had its objectives defined as: "Provide in industry a capability to undertake required engineering, maintenance and spares provision support for the aircraft, its systems, equipments and support facilities during the service life of the aircraft; establish, maintain or enhance the defence industry capabilities in general and provide a balanced stable ongoing workload using the opportunities presented by the project".

Offset work was an important adjunct to the programme and work to the value of at least 30 per cent of the imported content was sought. The preferred work was the production for export of components built in Australia for the new fighter.

In November 1978 the contenders for the TFF contract (MD, GD and Dassault the main ones at that stage) were invited to seek competitive bids from Australian industry on all tasks associated with the project except for airframe and engine final assembly and test which were already allocated to GAF and CAC, respectively.

By the time the Hornet was selected in October 1981, the industry programme had been defined and as the Australian participation programmes from both McDonnell Douglas and General Dynamics were considered satisfactory, the decision in favour of the Hornet came down to operational considerations. By the beginning of 1982 the necessary Deeds of Agreement between the Australian government, McDonnell Douglas and General Electric had been signed and the work of putting it all together could begin.

The Hornet's Hughes AN/APG-65 pulse-doppler multi mode digital radar, a key component of the aircraft's operational effectiveness.

THE AUSTRALIAN HORNET

The Australian Hornet manufacturing programme involved many local companies, the introduction of new technologies and considerable investment in plant and equipment.

The major companies involved and their roles are described below, but in the meantime it is worth mentioning some organisational changes which took place within the major players – GAF and CAC – after the project got underway.

The Commonwealth owned Government Aircraft Factory (which had previously built Beauforts, Beaufighters, Canberras, Lincolns, Mirages and Nomads) was reorganised and rationalised to become the much more efficient Aerospace Technologies of Australia (ASTA) in 1987 while the privately owned Commonwealth Aircraft Corporation was merged with Hawker de Havilland Ltd (headquartered at Bankstown NSW) to become Hawker de Havilland Victoria (HDHV) from 1986.

For convenience, these companies will from here mainly be referred to by their current names or at least abbreviations thereof – ASTA, HDH and HDHV.

The RAAF's Hornets featured some changes to the American specification to meet Australian requirements, including deletion of the catapult attachment points on the nosewheel leg, the incorporation of a landing light, standard RAAF undercarriage down tone, Australian Fatigue Data Analysis System (AFDAS), improved video/voice recorders, a landing light, standard Martin Baker ejection seat harness instead of the US Navy item, provision for a reconnaissance pod, clearance for the use of standard RAAF practice bombs, HF radio and replacement of the US Navy's carrier landing system with ILS/VOR for land based operations.

Two aircraft were fitted with extensive flight test instrumentation for trials work. (Subsequent modifications and upgrades will be discussed later.)

The standard armament for the RAAF's Hornets include AIM-9L Sidewinder 'all aspect' air-to-air missiles, AGM-88 anti radiation missiles, AGM-84 Harpoon anti shipping missiles and 2,000lb (907kg) laser guided 'smart' bombs.

The 75 Hornets equip three operational squadrons (Nos 3 and 77 at Williamtown NSW and No 75 at Tindal NT) and No 2 Operational Conversion Unit (OCU), also at Williamtown. The single seat F/A-18As were allocated the RAAF serial numbers A21-1 to -57 and the two seat F/A-18Bs A21-101 to -118. The first two aircraft (A21-101 and 102) were assembled and flown in the USA and the remainder in Australia at ASTA's Avalon, Victoria, facility.

The production and delivery schedule called for the handover of two aircraft in 1984 (the pair of US built two seaters), five in 1985, 15 in 1986, 18 in 1987, 18 in 1988, 13 in 1989 and four in 1990, the last of them to be presented to the RAAF in May of that year. The latter date was successfully achieved although deliveries fell to about eight units behind schedule during 1987. This was recovered and by the end of 1988 the programme was back on target.

The RAAF's first two Hornets (A21-101 and 102) overfly Sydney Harbour after their 15 hour non stop delivery flight from California to Australia. Both carry the tiger's head motif of No 2 Operational Conversion Unit on their fins.

An RAAF 75 Sqn Hornet flies wing on an Su-27 Flanker during Asian Aerospace. (Katsuhiko Tokunaga)

Two early RAAF F/A-18B Hornets: A21-103 (top) at St Louis in April 1984 before disassembly and transport to Australia and A21-101 (bottom) flown non stop from California to Williamtown in May 1985 in company with A21-102. (MDC)

Hornet A21-114 of No 77 Squadron (top) at RAAF Fairbairn in October 1990 with a No 1 Squadron F-111C behind. The last Australian Hornet, A21-57 of No 3 Squadron (bottom) flies over Stockton Beach near Newcastle NSW wearing a special 75th anniversary colour scheme. (RAAF)

Hornets several generations apart (top) – the McDonnell Douglas variety and the de Havilland Hornet Moth of 1935. In its element, F/A-18A Hornet A21-9 (bottom) banks over the Australian scrub.

3 Squadron Hornets at low level: A21-16 (top) over the coast of New South Wales and A21-57 over RAAF Point Cook with CT-4 Airtrainers visible on the ground.

Williamtown scenes: F/A-18B Hornet of No 2 OCU in July 1990 (top) and F/A-18A A21-10 of No 3 Squadron. Note the leading edge extension strakes on both aircraft. (Bruce Chapman)

Works Programmes

The RAAF Hornet programme required a massive investment in capital works and machinery by the Department of Defence. By 1985, $27m had been spent on 24 capital works projects, and 356 separate items of machinery valued at $88m had been acquired.

Highlights included the establishment of an avionics test facility at Avalon, a chemical milling plant at ASTA's Fisherman's Bend (Vic) facility, a flight test shelter and fire protection at Avalon and an engine test facility at HDHV's Fisherman's Bend plant.

Machinery which had to be acquired included a honeycomb profiler, a new autoclave for composite materials, an NC vertical lathe, five axis machining centres and vertical heat treatment equipment.

An $8m chemical milling plant at ASTA was the centre of some trouble in the early days. The company removed the plant from its list of requirements in 1981 after HDHV had successfully tendered to produce engine doors and aft fairings under sub contract to Northrop and it was assumed the latter would install a chemical milling plant.

The government's defence support rules and regulations then came into play, namely that new plant and facilities would be installed where similar facilities already existed and ASTA had a small facility. Some six months were lost while this was sorted out and a further four months while the finances for the plant awaited approval. The delays also affected HDHV, of course, which had to buy parts sets from Northrop for a time.

The plant was commissioned in 1984 and schedules caught up, but not before industrial troubles affected other plans. The building of a $25m structural bonding plant at ASTA was delayed when the Builders Labourers' Federation (one of the most militant, troublesome, corrupt and belligerent unions in the history of the industrialised world) argued with the contractor (Civil and

Two views of the first Australian assembled Hornet (A21-103) at the McDonnell Douglas St Louis plant. The aircraft was then disassembled, flown to Australia in the hold of a C-5 Galaxy transport and reassembled by ASTA at Avalon. (MDC)

Civic) over the use of scissor lifts! But this was the BLF, and any excuse to cause trouble was standard procedure ...

All items of plant and equipment were purchased after a rigorous contractual process which, according to R M Stainer, Director, Tactical Fighter Force Project Office, "... demonstrated to those involved the rise of the Japanese machine tool industry, and the dismal state of the Australian industry".

Despite this, there were substantial benefits for Australia as a whole as all this equipment could be used not only for the Hornet programme but also for other projects (defence or commercial) which may follow.

Australian Industry Participation

After all the contracts had been let and the deals done, Australian industry involvement in the Hornet programme involved numerous companies which produced a large part of the Hornet.

Local contributions to the Australian Hornet programme are summarised in the table below:

AUSTRALIAN HORNET CO-PRODUCERS

Aerospace Technologies of Australia (GAF)
- Aircraft final assembly and test
- Forward fuselage installations
- Trailing edge flap assembly
- Trailing edge flap shroud assembly
- Canopy transparency
- Canopy assembly
- Radome assembly
- Windshield transparency

Hawker de Havilland Victoria (CAC)
- Wing pylons
- Engine access door assembly
- Aft nozzle fairings assembly
- Aircraft mounted accessory drive gearbox
- General Electric F404 engine

Hawker de Havilland
- Landing gear
- Stabilator actuator
- Leading edge flap servo valve
- Aileron actuator
- Rudder actuator
- Trailing edge flap actuator
- Hydraulic reservoir

Dunlop Aviation Australia
- Nosewheel assembly
- Main wheel and brake assembly
- Speedbrake actuator

Philips Electronics Australia
- Hughes APG-65 radar
- Radar data main processor

Morris Productions
- Multilayer printed circuit boards

Computer Sciences Australia
- Automatic Test Equipment (ATE) software

Hornet A21-103 undergoing assembly in the USA before being dismantled and flown to Australia in a USAF Galaxy for reassembly. The Hornet next to it is destined for Canada. (MDC)

British Aerospace Australia
Avionics fault tree analyser
Up front control panel
Head Up Display
Multipurpose display group
Flight control computer
Normalair-Garrett Australia
Auxiliary power unit
Air conditioning system components
Aircraft Equipment Overhaul Sales
Hydraulic pump assembly
Thorn EMI Electronics Australia
Inertial navigation system
Stores management system
Air data computer power supply
Communications system control set
Fire and bleed air detection system
Lucas Industries Australia
Variable speed constant frequency unit
Main fuel control unit
Afterburner fuel control unit
Standard Telephone and Cables (STC)
Flight control computer power supply

Hawker de Havilland Victoria (CAC) has a long pedigree in the manufacture of both engines (Wasp, Twin Wasp, Merlin, Nene, Avon, Viper, Atar) and airframes (Wirraway, Wackett, Boomerang, Mustang, Winjeel, Avon-Sabre, Mirage wings, Macchi) in Australia and as such was well placed to be a major participant in the Hornet programme.

Its contribution to the Hornet airframe was relatively minor compared with previous programmes, but its expertise in engines continued with the General Electric F404.

The company was involved in the production of 158 F404s plus parts of a further 17 engines were assembled from modules. The first seven engines were supplied by General Electric.

Apart from assembling the engines, HDHV was responsible for the manufacture of 27 components totalling about 100 parts. The company had originally bid for the manufacture of 52 components but it was decided that the cost of tooling outweighed the benefits to be derived from technology transfer.

Considerable investment was needed to establish the new technology at HDHV, as it was elsewhere, but considering the overall object of the Australian Industry Participation programme – to bring Australia to the forefront of modern technology – the cost was modest.

An $8.7m engine test cell was built for the F404 and is capable of testing up to five engines per month (assuming one shift per day) including engines returned from service for depot level maintenance and repair. Capable of accommodating engines of up to 40,000lb thrust, the cell has obvious potential for engines other than the F404.

Other equipment installed at HDHV for the engine programme include an electrochemical grinder, a three axis numerically controlled machine, blade grinders, compressors, turbine assemblies and a borescope inspection machine. New engine repair equipment included an electron beam welder, a vacuum brazing and heat treatment machine and high temperature diffusion coating machine plus welding and ultrasonic immersion equipment and an electron scanning microscope.

HDHV's Hornet airframe activities also required new equipment, including a high pressure bladder press forming machine with a capacity of 14,000psi, titanium hot forming equipment, numerically controlled turrets, welders, grinders, cutters and jig borers and so on. The investment was not wasted. When all engines had to be quickly modified to protect the aircraft from titanium fires in the event of a blade failure, HDHV was the first company in the world capable of performing the work. All the RAAF's F404 engines were modified locally.

An interesting sidelight gives some idea of the logistics involved in the Hornet programme. Engine bay doors and nozzle fairings built by HDHV were first shipped to Northrop in California where they were fitted to rear fuselages. Then the fuselages were sent to Australia for fitout and final assembly!

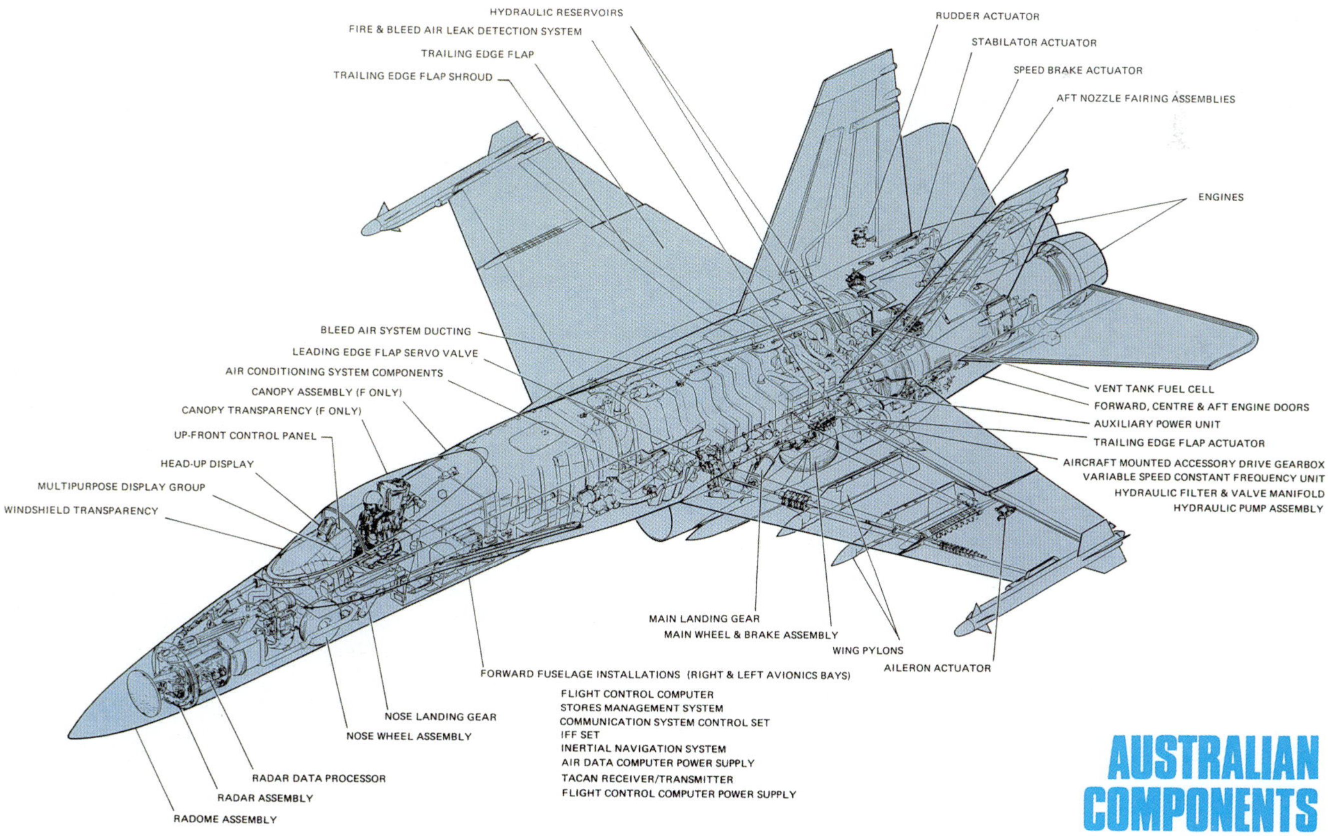

AUSTRALIAN COMPONENTS

The first RAAF single seat Hornet – A21-1, in 2 OCU markings – in company with the aircraft it replaced in RAAF service, the Mirage IIIO. A21-1 was delivered in November 1985. (DoD)

An RAAF F/A-18B Hornet at the McDonnell Douglas plant – complete but unpainted. (MDC)

A21-103 about to be swallowed up by the USAF Galaxy which took it to Australia in June 1984. After reassembly, it was flown at Avalon on 26 February 1985.

An Australian assembled Hornet is given its 'shower test' – for water proofing – one of the numerous tests each aircraft underwent before delivery.

A big moment. A21-103 – the first Australian assembled Hornet – is formally rolled out with due ceremony at Avalon on 16 November 1984.

An even bigger moment – A21-103 records its first flight at Avalon on 26 February 1985 in the hands of McDonnell Douglas pilot Rudi Haug. The aircraft reached Mach 1.6 and an altitude of 40,000 feet during the flight.

Hornet A21-103 'cleans up' after takeoff on an early test flight.

ASTA's most visible contribution to the Hornet programme was its final assembly work at Avalon airfield, which is now owned by the company. The coming together of local and imported components and the complex fitting out of the airframe with the myriad of electronics and other equipment took about 12 months to complete for each aircraft.

New technology brought new equipment to ASTA – a new avionics building at Avalon, a bonding plant at Fisherman's Bend costing $25m and featuring state-of-the-art machinery and autoclaves for bonding of graphite/epoxy material used in the Hornet, the chemical milling plant mentioned earlier, a five axis honeycomb profiler for machining the core of the flaps, an ultrasonic scanner for non destructive testing of the flap shrouds and so on, all of them representing "a major quantum jump in technology", according to the GAF/ASTA F/A-18 project manager, Brian McCarthy.

The Sydney arm of Hawker de Havilland (HDH) had perhaps the most diverse involvement in the Hornet programme providing and/or assembling parts under sub contract to no fewer than 10 American companies. The main and nose undercarriages are the most obvious components but equally important are control surface actuators for the stabilator, ailerons, rudder and flaps.

Again, new technology and techniques were the order of the day due to the fly-by-wire controls of the aircraft and there was of course considerable investment in new equipment – three axis three spindle and four axis single spindle numerically controlled machines, a vacuum heat treatment oil quench furnace and a tally centre machine which automatically aligns components and computes their true shape.

The importance of the radar and avionics systems in the Hornet are obvious – it's very much an electronic aeroplane – and the companies involved in these aspects of the programme were very much at the leading edge of the technology.

Philips Electronic Services teamed with Hughes for the Hornet's AN/APG-65 radar, a compact and versatile multi mode radar which weighs only 339lb (154kg). The AN/APG-65 comprises five major components – programmable signal processor, radar data processor, receiver-exciter, transmitter and antenna. Philips manufactured the radar data processor (RDP) and assembled all the radars for the RAAF's Hornets.

Before the Hornet programme there was no such thing as a military radar support industry in Australia but Philips has been able to build on its expertise and establish an internationally recognised facility which provides support for the RAAF as well as setting the scene for participation in other programmes.

Other areas which have resulted in ongoing benefits for local industry include the British Aerospace Australia designed and built Avionics Fault Tree Analyser (AFTA) which has been exported to Canada and Spain.

The Hardware Flies

With the industrial and commercial aspects of the Hornet programme in place, it was time to start the process of actually building the aircraft and delivering them to the RAAF.

Ground crew members pay close attention to the RAAF's first Hornet at Avalon.

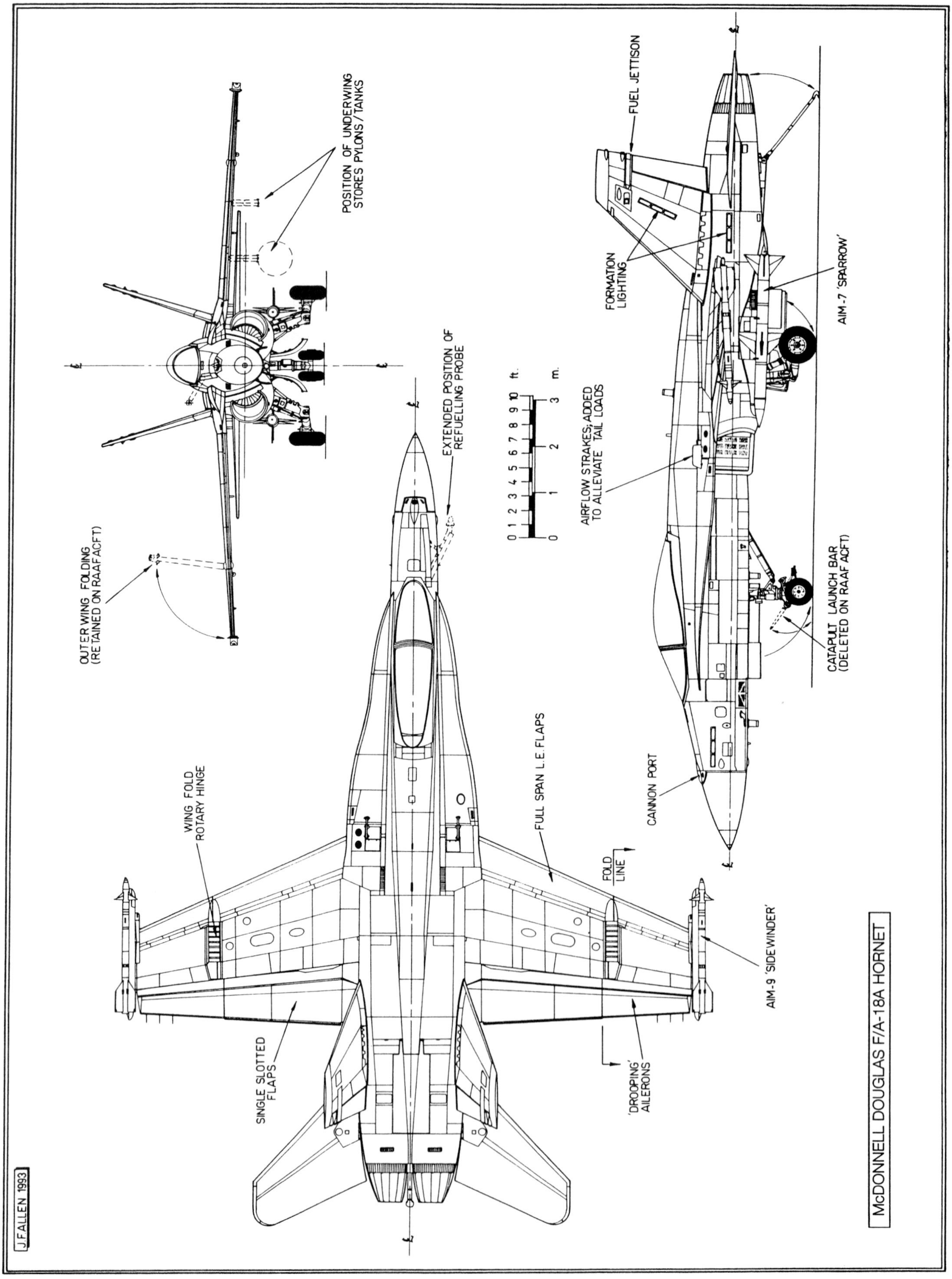

McDONNELL DOUGLAS F/A-18A HORNET

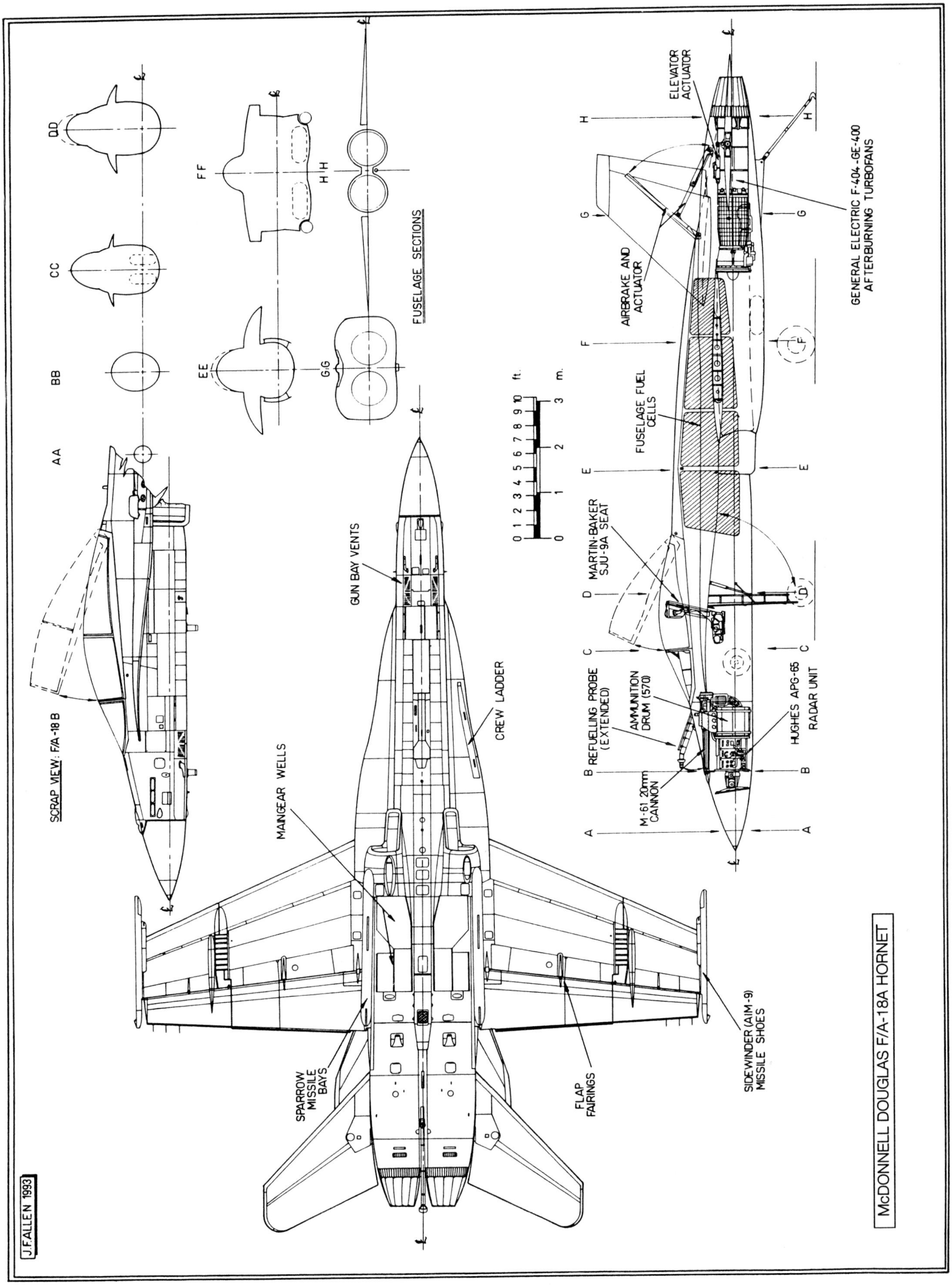
McDONNELL DOUGLAS F/A-18A HORNET
J.F.ALLEN 1993
SCRAP VIEW: F/A-18 B
FUSELAGE SECTIONS
AA
BB
CC
DD
EE
FF
GG
HH
MAINGEAR WELLS
GUN BAY VENTS
CREW LADDER
SPARROW MISSILE BAYS
FLAP FAIRINGS
SIDEWINDER (AIM-9) MISSILE SHOES
REFUELLING PROBE (EXTENDED)
AMMUNITION DRUM (570)
M-61 20mm CANNON
HUGHES APG-65 RADAR UNIT
MARTIN-BAKER SJU-9A SEAT
FUSELAGE FUEL CELLS
AIRBRAKE AND ACTUATOR
ELEVATOR ACTUATOR
GENERAL ELECTRIC F-404-GE-400 AFTERBURNING TURBOFANS
ft.
m.

Hornet A21-103 over RAAF Point Cook.

In order to achieve the planned delivery schedule, a production rate of about 1.5 aircraft per month had to be achieved once things were in full swing. The first two aircraft (RAAF serials A21-101 and -102, ASTA constructor's numbers ATF-1 and -2) would be built and flown in the USA and later delivered to Australia. A21-103 (c/n ATF-3) would also be bolted together in the USA but then dismantled, transported to Avalon by air and reassembled locally, in company with A21-104 (c/n ATF-4).

A21-103 would arrive complete but in pieces whereas A21-104 would be delivered *sans* engines, undercarriage and avionics and would therefore be regarded as the first Australian 'manufactured' Hornet. Of the first eight RAAF Hornets (scheduled to be delivered in 1985) seven would be two seater F/A-18Bs and the eighth a single seater F/A-18A.

The first RAAF Hornet, A21-101, recorded its maiden flight at St Louis on 13 August 1984. A21-102 followed shortly afterwards and both were formally handed over to the RAAF by McDonnell Douglas on 29 October. These aircraft remained in the USA until May 1985 when they were flown non stop from US Naval Air Station Lemoore, California, to Williamtown NSW, arriving on the 17th of that month to be greeted by a large crowd of officials, the press and other interested parties.

The two Hornets were accompanied on the 15 hour, 6,830nm (12,640km) flight by a USAF McDonnell Douglas KC-10 Extender tanker which refuelled the Hornets 13 times *en route*. The KC-10 itself had to be refuelled by another Extender over Hawaii. The Hornets were flown by Wng Cdr Brian Robinson (CO of No 2 OCU) and Flt Lt A Larard in A21-101 and Sqdn Ldr Laurie Evans and Flt Lt G O'Brien in A21-102.

Rollout and First Flight Delays

In the meantime, the first two Australian assembled/manufactured Hornets (A21-103 and -104) had been flown following their arrival at Avalon on 6 June 1984 in the hold of a US Air Force Lockheed C-5A Galaxy.

On 16 November 1984, A21-103 was formally rolled out at Avalon in the presence of Prime Minister R J Hawke, the Chief of Air Staff, Air Marshall David Evans, and a gathering of guests from the government, the Department of Defence, the US Navy, McDonnell Douglas and Australian industry representatives. On 28 November the first HDHV/CAC assembled F404 engine was handed over.

The first flight of A21-103 was originally scheduled for the same month, but a demarkation dispute over who should be at the controls of the aircraft on that important day delayed things.

The matter had interesting legal implications (who is legally allowed to fly a military aircraft not yet owned by the Australian government in Australian airspace), operational implications and industrial implications, the Australian Federation of Air pilots (who five years later would provoke the notorious pilots' dispute which would cripple Australian domestic air services as well as destroy the union's credibility along with the careers of most of its members) demanding its members fly the aircraft.

The issue was finally sorted out when arrangements were made for contractor test pilots to perform the first (contractor's) flights and US Navy pilots or RAAF officers seconded to NAVPRO to fly the customer test flights. Customer acceptance flights were performed by RAAF pilots.

The result of all this was that McDonnell Douglas pilot Rudi Haug took A21-103 up for its maiden flight on 26 February 1985. An attempt to record the first flight was actually made the previous day, but it was abandoned due to unfavourable weather conditions while the Hornet was taxying out to the runway.

Haug flew for 66 minutes on 26 February and recorded a second, 90 minute flight on 7 March. Haug described the flights to aviation journalist Gerry Carman and subsequently published in *Aircraft* magazine: "The normal test profile is to have a good look at the systems in the aeroplane; the flight worthiness and safety of the airframe, flight controls, engines and life support system.

"The profile began with systems checks on the ground. During 15-20 minutes I checked about 100 items before hitting the runway. I selected full power, afterburner, for takeoff. I immediately climbed to more than 40,000 feet, checking the environmental controls, fuel, trim and general flight worthiness during the climb. Lining up at high altitude I initiated specific tests ... very low speeds below 200 knots at high altitude, engine checks, individually and then combined and then a high speed supersonic dash (about Mach 1.6) at around 40,000 feet.

"During the speed run I checked fuel transfer, handling qualities and cabin environment systems. At the end of the run there were airframe to engine integration checks with rapid deceleration using speed brakes.

"Then, a 'wind up' turn to put some high 'Gs' from +7g -1g on the airplane for structural integrity and control checks and into descent in idle power, checking cabin pressure and fuel flow. Then each engine was shut down and restarted separately.

"Next, we monitored the navigation aids, UHF and VHF, and some high angles of attack including stalls, checking the variable lift devices – leading and trailing edges ..." and so on, including checks of the undercarriage retraction/extension, the aerial refuelling system and partial avionics tests.

The second flight included tests of the AN/APG-65 radar – its integration with the avionics and simulated weapons selection, lock on and launch using a Macchi chase aircraft as the 'target'.

There was one more formal ceremony to be performed before the Hornet's RAAF service got underway. This was held at Avalon on 4 May 1985 when Chief of Air Staff Air Marshall David Evans was handed A21-103's documents by United States *Charge d'Affaires* Mr David Lambertson. Guests included Defence Minister Kim Beasley, GAF general manager Ray Pascoe and McDonnell Douglas senior vice-president Alvin Boyd.

RAAF test pilot Sqdn Ldr Mark Hayler flew an impressive display routine in the aircraft.

Production – Problems and Solutions

The rate of delivery of Hornets to the RAAF can be seen in the accompanying "Summary of Annual Deliveries" table. All 18 F/A-18Bs had been delivered by the end of 1988 and the last RAAF Hornet (A21-57) was handed over on schedule in May 1990.

There were some delays in the meantime, but these were recovered. With the delivery in September 1988 of A21-115, the RAAF's 53rd Hornet, the production programme was back on track after an eight aircraft shortfall at the time the Government Aircraft Factory was reconstituted as Aerospace Technologies of Australia in mid 1987.

In the first half of 1987, GAF produced only one Hornet per month instead of the required 1.5, and restoration of the planned programme required a 100 per cent increase in production for a time. For its part, GAF claimed work on Hornet flaps and flap shrouds was running late due to faulty jigs being received from McDonnell Douglas.

Investigations tended to indicate that GAF's general inefficiencies (most of them resulting from union dominance and therefore work practices) were to blame, and in fact all of GAF's projects were behind schedule. Hornet components which should have been locally made had to be air freighted from McDonnell Douglas at considerable cost and naturally, Hornet final assembly and delivery slowed down.

Assembly scene at Avalon with A21-107 in the foreground. This F/A-18B was handed over to the RAAF in December 1985.

The much more efficient ASTA was able to recover the situation using 100 fewer workers than GAF had employed on the Hornet programme!

By the end of 1988, 59 Hornets had been delivered to the RAAF and the programme was back on schedule. The final example (A21-57) was handed over to the RAAF at Fairbairn (Canberra) ACT on 16 May 1990, following a flight from Avalon by Grp Capt Ray Conroy. Escorted by four other Hornets, Conroy's arrival was spectacular, a full throttle vertical climb being followed by a display by the escorting Hornets and ending with a low level pass and landing by A21-57. Conroy then presented the handover documents, the RAAF Command Band played the Australian and US national anthems and the 300 invited guests looked on.

The delivery phase of the Australian Hornet programme was over.

Mr James L Spehr, Vice President General Manager F/A-18 Programme for McDonnell Douglas reviewed the project at the handover: "On this, the occasion of the delivery of the 75th McDonnell Douglas F/A-18 Hornet to the RAAF, we welcome you and invite you to share our pride in the achievements of the industry team that enabled this event to be what it is. An on time, on cost delivery of the final Hornet in the Australian programme – a nine year enterprise.

"Together with our industry partners, here and in the US, we are pleased that this highly capable aircraft has joined one of the world's most professional air forces and that the RAAF is finding it to be an outstanding fleet addition – helping to meet the diverse operational requirements of a large and isolated nation.

"This event is another milestone in the long standing partnership between McDonnell Douglas and the Australian flying community – both military and civil, The Hornet is destined to remain the cornerstone of Australia's tactical fighter force well into the next century. As technology developments make the F/A-18 even more capable, we look forward to a continuing relationship with both Australian industry and the RAAF in achieving the modifications which those developments will precipitate".

RAAF HORNETS
SUMMARY OF ANNUAL DELIVERIES

	18A	18B	Total
1985	1	7	8
1986	12	–	12
1987	7	7	14
1988	21	4	25
1989	12	–	12
1990	4	–	4
TOTALS	**57**	**18**	**75**

F/A-18A A21-8 over Point Cook wearing the markings of No 3 Squadron, the first RAAF fighter squadron to re-equip with the Hornet. (DoD)

RAAF Hornets regularly deploy to Butterworth in Malaysia, 2 OCU's A21-102 (top) is photographed here in the area. Nature turns on a spectacular backdrop for 75 Squadron's A21-43 (bottom).

Reach for the sky! A pair of 3 Squadron Hornets demonstrate their rates of climb.

Force multiplier. The RAAF's Boeing 707 tankers have made the Hornets an even more useful aircraft, enabling distant deployments to be made in a short time.

Hornet trainers on parade: F/A-18Bs A21-116 (top) of 75 Squadron and A21-110 (bottom) of 2 OCU, photographed at Darwin in May 1989 and July 1990, respectively. (Bill Lines)

Hornets exercising at Darwin in 1990, No 77 Squadron's A21-48 (top) photographed from a Boeing 707 tanker, while aircraft from 3 Squadron taxy out to begin a sortie. (Bill Lines)

HORNET IN SERVICE

The three RAAF Hornet squadrons – Nos 3 and 77 at Williamtown NSW and No 75 at Tindal NT – carry between them the responsibility of providing Australia with the capability of performing varied roles in the defence of the realm, roles which would provide the nation with a military capability not only to discourage threats but also to be able to counter – unaided – enemy action if necessary, at least for a time.

The versatile F/A-18 Hornet provides the RAAF with that kind of capability, in both the air-to-air and air-to-ground roles. In Australian service the aircraft is expected to intercept intruding aircraft and to dogfight plus provide interdiction of lines and communication and supply, and perform maritime strike, plus close air support for the Army.

The introduction of the Hornet to RAAF service brought with it considerable expenditure on bases and facilities, most of it as a direct result of the aircraft's entry to service and some as part of general defence upgradings.

The main Hornet base at Williamtown had over $150m spent on the upgrading of the runway and taxiways, new hangars and squadron buildings and maintenance facilities.

RAAF Tindal, near Katherine NT and some 260 kilometres south-east of Darwin, is a new facility for 75 Squadron's Hornets. The former wartime strip began its transformation into a modern combat aircraft complex in 1985 and was completed in 1988 at a cost of $215m. The establishment of Tindal as a fighter base removed Australia's first line of defence from its traditional base at Darwin, providing a little more reaction time and an easier operating environment but with no loss of operational effectiveness due to the installation of Jindalee over the horizon radar.

Two other RAAF bases in northern Australia have come about since the Hornet acquisition was announced. These 'bare bones' bases provide runways, taxiways, parking space and refuelling services but not the large number of buildings and other facilities associated with a major base.

The idea is to provide a base which can be quickly used for operational purposes if required and is normally staffed by a skeleton crew. The first of them, RAAF Curtin, near Derby in Western Australia, was opened by Prime Minister Bob Hawke in June 1988, while work on a second such base (RAAF Scherger near Weipa, Queensland, on Cape York Peninsula) was scheduled to begin in 1993 for completion in 1997-98.

These bases are and will be used for exercises designed to test the RAAF's capability to operate from a sparse environment.

One exercise at RAAF Curtin involved Hornets from No 75 Squadron (based at Tindal) and Northrop F-5E and General Dynamics F-16 fighters from the Republic of Singapore Air Force together with two KC-130 Hercules tankers and an E-2C Hawkeye early warning aircraft. In addition, RAAF F-111s, a Boeing 707 tanker for the Hornets and a pair of RAAF Air Defence Radar Units also took part.

Exercise Western Reward was designed as an air defence exercise but included a ground war scenario involving the Army's 13th Brigade and the RAAF's No 2 Airfield Defence Squadron. Their role was to protect the airfield on the ground.

In order to bring the base up to a full operational standard, 56 truckloads of equipment were transported to Curtin from all over Australia.

A change to a traditional RAAF role coincided with the arrival of the Hornet in RAAF service and the retirement of the last operational Mirages. For three decades, the RAAF had maintained a permanent pres-

No 3 Squadron was the first RAAF squadron to equip with Hornets, the first aircraft arriving in August 1986. The squadron's A21-8 and A21-18 formate here, the former wearing a shark's mouth design on its nose. (Defence PR)

F/A-18B A21-102 of 2 Operational Conversion Unit eases up towards the camera.

The three Williamtown based Hornet units are represented in this shot, from top to bottom: A21-6 of 2 OCU, A21-9 of 3 Squadron and A21-4 of 77 Squadron. (Defence PR)

ence at Butterworth, Malaysia, which at its peak housed some 2,000 Australian servicemen and women and their families plus more than 1,000 local civilians who worked on the base.

RAAF Sabres and then Mirages were stationed at Butterworth, but a change of policy saw the end of permanent RAAF residency there in April 1988, 75 Squadron returning to Australia to swap its Mirages for Hornets after a deployment which lasted 16 years. This was not the end of the RAAF's involvement with the Malaysia/Singapore region, however, as with the announcement of the RAAF's withdrawal from Butterworth came the commitment to deploy Hornets to the base for a total of at least 16 weeks each year. All three current RAAF Hornet squadrons are regular visitors to Butterworth, along with F-111s, Orions and other aircraft.

The RAAF's Hornet squadrons come under the control of the Tactical Fighter Group, which is commanded by the Officer Commanding RAAF Williamtown. The Group includes the Tindal based Hornets as well as those at Williamtown (including the OCU) plus Nos 25 (Pearce WA) and 76 (Williamtown) Squadrons with Macchi MB.326 jet trainers for fighter lead-in work. No 81 Wing was formed to control Williamtown's Hornets while No 481 (Maintenance) Squadron at Williamtown is responsible for maintaining and repairing the Hornets as well as training all technical personnel who will eventually work on the fighter.

Hornet Squadrons

All three RAAF Hornet squadrons have long traditions and numerous battle honours behind them.

The first to receive Hornets was No 3 Squadron, which took delivery of F/A-18As A21-8 and A21-9 on 29 August 1986 after a 55 minute flight from Avalon to Williamtown. No 77 Squadron was next, receiving its first Hornet on 29 June 1987, while No 75 Squadron was the last to begin re-equipment, in May 1988. At that stage 75 Squadron resided in Darwin and moved to Tindal in early October.

No 3 Squadron (motto: *Operta Aperta* – 'Always On Guard') has its origins as an Australian Flying Corps (AFC) squadron formed at Point Cook, Victoria, in 1916. It was the first Australian squadron to serve on the Western Front during World War I (as No 69 Australian Squadron, Royal Flying Corps) before reverting to its AFC identity in 1918.

Disbanded in 1919, the squadron reformed as an RAAF unit in 1925, flying DH.9s, Westland Wapatis and Hawker Demons in the inter-war years. With the outbreak of World War II it was sent to the Middle East, serving as an army co-operation squadron in Libya (with Lysanders) before becoming a fighter squadron under RAF control equipped with Tomahawks, Kittyhawks and then Mustangs, which it operated in Italy carrying the now familiar 'southern cross' tail markings.

No 3 Squadron was disbanded in 1946, lived on as No 3 (Tactical Reconnaissance) Squadron between 1948 and 1953 (from the renamed 4 Squadron) reformed again in 1956 as a fully operational unit equipped

RAAF Hornets A21-101 and -102 were flown non stop from California to Williamtown on their delivery flight in May 1985. They were accompanied by a USAF KC-10 tanker and escorted into Williamtown by a pair of Mirages. Approaching Williamtown, A21-102 is in company with 77 Squadron Mirage A3-95. (RAAF Historical)

A 75 Squadron Hornet (A21-116) taxies into its protective shelter at RAAF Tindal in the Northern Territory. (Defence PR)

Some good underside detail of 3 Squadron's A21-8 as it turns away from the camera. (Defence PR)

The latest and the greatest? No 3 Squadron Hornet A21-12 poses with Dakota A65-86 of the Aircraft Research and Development Unit (ARDU) at RAAF Fairbairn, Canberra. (RAAF)

Hornets from 2 OCU in company with C-130E Hercules A97-190 from No 37 Squadron.

3 Squadron Hornets A21-8 and -18 pose for the camera. (Defence PR)

Who'd have thought it possible not so long ago? Hornets A21-49 and -106 formate with a pair of Soviet Sukhoi Su-27 Flankers during the Singapore Airshow in early 1990. Note the strakes fitted to the Hornets' leading edge extensions, a modification designed to lessen the damaging effects of certain airflow conditions around the base of the fins. (Defence PR)

with Sabres and then Mirages. It spent most of this period at Butterworth before re-equipping with Hornets.

No 77 Squadron ('Swift to Destroy') was formed at Pearce WA in March 1942, and equipped with Kittyhawks provided air defence for the Perth area before being deployed to Darwin the following August where it saw action against the Japanese. The action continued throughout the war, its Kittyhawks flying from various bases in the South West Pacific Area including Milne Bay, Goodenough Island and Morotai.

By the end of the war 77 Squadron was at Labuan, Borneo, where it re-equipped with Mustangs. With these aircraft it moved to Japan in 1946 as part of the British Commonwealth Occupation Force but after the decision had been made to return the squadron to Australia in 1950 the Korean war began and No 77 found itself in action again, flying Mustangs and then Meteors in that conflict.

The squadron finally returned to Australia in 1954 and flew Meteors from Williamtown before re-equipping with Sabres in late 1956. It was deployed to Butterworth in 1959 and remained there until 1969 when it returned to Williamtown, by now equipped with Mirages.

No 75 Squadron ('Seek and Destroy') was another unit which was formed when Japan's actions threatened Australia. Established at Townsville Qld in March 1942 and equipped with Kittyhawks, it found itself in the thick of the action just 17 days later in the desperate defence of Port Moresby.

75 Squadron flew Kittyhawks throughout the war, operating in the New Guinea area (including at Milne Bay) and from the islands to the north and finally Borneo as Japan's war effort crumbled. The squadron re-equipped with Mustangs in 1946 and operated them from Williamtown for two years until it was disbanded.

Reformed at Williamtown in early 1949 and equipped with Mustangs and then Vampires, the squadron then spent over two years from mid 1952 in Malta flying RAF Vampires as part of the Mediterranean air defence, but on its return to Australia it became the RAAF's first Sabre unit in April 1955, flying from Williamtown. The squadron was also the first to be equipped with Mirages (August 1965) and the first Mirage squadron to be deployed to Butterworth (May 1967).

No 75 Squadron returned to Australia in 1983 remaining at Darwin until its Mirages were replaced with Hornets in 1988.

2 Operational Conversion Unit

The first RAAF unit to receive Hornets was No 2 OCU at Williamtown, scene of Hornet conversion training. 2 OCU had previously been responsible for Mirage conversion training, but this role was handed over to 77 Squadron in January 1985 while the OCU geared up for the arrival of the Hornet.

For a time, 77 Squadron had the largest peacetime complement of any RAAF squadron with some 40 Mirages and 16 Macchis (also ex 2 OCU) on strength, taking care of not only its regular operational duties but also the training of the last batches of new Mirage pilots. The squadron completed the final Mirage Conversion Course (No 47) in February 1986.

2 OCU began its first Hornet conversion course in August 1985 with the RAAF's then entire Hornet fleet on strength – F/A-18Bs A21-101, 102, 103 and 104 – along with just three students, two of them experienced Mirage pilots who would themselves become instructors. The first six instructors were trained by the US Navy's first Hornet squadron, VFA-125 at Leemore, California.

As the number of Hornets on strength increased, so did the momentum of the courses and the number of pilots passing through it. In 1987, 21 pilots passed

RAAF Hornets A21-101 and -102 show themselves off to the Australian public over Sydney (top) and Perth (bottom). Both are carrying the markings of 2 OCU.

through the OCU and four different courses were being run – two advanced courses for pilots with previous fighter experience and two basic courses for new pilots direct from flying training school and pilots who had moved to fighters from other types such as transports or helicopters.

Fourteen of the 18 Hornet two seaters had been delivered by the end of 1987 and the remainder came on line the following year. Most were with 2 OCU but others were with the squadrons and with the ARDU.

In association with its aircraft, 2 OCU naturally has state-of-the-art training equipment at its disposal including an advanced computer controlled flight simulator which is capable of reproducing the Hornet's entire flight envelope and a device called the HOTASTA, or Hands On Throttle And Stick Training Aid. With most of the operational switches mounted on either the Hornet's throttle levers or stick (so the pilot's hands don't have to leave those essential controls during combat) this is an important item because it allows students to become fully familiar with them before flying the real aircraft.

Computers are also extensively used for the ground instruction courses and incorporate visual display units (VDUs) allowing the students to progress at a pace suitable for them and for them to be able to easily revise what they've been taught.

Fighter Combat Instructors

Another vitally important element of 2 OCU's activities is that of conducting Advanced Fighter Combat Instructors' courses. These are held every two years and run for 18 tough weeks, the result being a pilot with advanced skills and knowledge of all the aspects of combat flying, from planning missions through to the uses of the various weapons available to a Hornet pilot.

Pilots selected for this intensive course come from the squadrons where they must have displayed above average skills in both air-to-air and air-to-ground combat. On completion of the course they either remain at the OCU as instructors or return to the squadrons where their expertise is put to operational use.

The final exercise the undergraduates from this course have to perform before being given their coveted Fighter Combat Instructor (FCI) badges is a true test of their skills. A typical mission might be like the one held in late 1992 when five students had to destroy two enemy targets, fight off up to 20 hostile aircraft and refuel from one of the RAAF's Boeing 707 tankers.

After departing Williamtown with their instructors in other Hornets, the handful of would-be FCIs set out to bomb the Evans Head bombing range where a simulated enemy air defence system of surface-to-air missiles lay in wait. They are then 'attacked' by other Hornets before going on to RAAF Amberley Qld for the night.

At Amberley, planning of the second phase of the exercise takes place. Departure from Amberley is set at 0300 hours the next morning and on the way back they will bomb the Singleton army range while facing attempted interception by no fewer than 16 'hostile' Hornets before carrying on to Williamtown and a celebratory glass of champagne to mark the end of the course.

Showing the Flag

2 OCU's Hornets do not remain firmly entrenched at Williamtown, they travel around to a large extent. This was particularly the case in the first year of RAAF Hornet operations when the OCU was the only unit operating the type and therefore was responsible for showing off Australia's new fighter.

Visits were made to every Australian state and territory, giving public demonstrations and staging deployments for certain phases of the training programme. Hornets deployed to RAAF Pearce, near Perth WA, and on the return trip to Williamtown flew the more than 2,000 miles (3,300km) non stop and unrefuelled.

Other journeys undertaken by 2 OCU's Hornets in the first year included visits to Singapore and Malaysia via Indonesia, the first overseas journey undertaken by RAAF Hornets. The trip included taking part in Singapore's Asian Aerospace '86 airshow, where the Hornet was a major exhibit.

A 75 Squadron Hornet at Tindal, this shot emphasising the very large flap area, helped by the drooping ailerons.(Defence PR)

In July 1990, 2 OCU's Hornets were temporarily displaced from Williamtown while that base's runway underwent major resurfacing. In order to allow training to continue, the 12 Hornets were deployed to RAAF Richmond west of Sydney, where they shared the facilities normally used by Hercules and Caribou transports.

The first operational overseas deployment of RAAF Hornets occurred in March and April 1987 when six aircraft from No 3 Squadron (A21-8, 10, 11, 12, 14 and 15) flew non stop to New Zealand to take part in the RNZAF's Golden Jubilee celebrations (which included air show appearances) and Exercise Willoh 87-1. It's interesting to note that three 77 Squadron Mirages flew to New Zealand at the same time, but they needed to refuel at Norfolk Island *en route*.

Force Multiplier

The New Zealand trip mentioned above demonstrates how much more readily deployable the Hornet is compared to the Mirage, but even greater flexibility is now available to the RAAF's fighter squadrons with the introduction of Boeing 707 tankers.

The RAAF has five 707s in service in 1993, four tanker/transports (A20-623, 624, 627 and 629) and a single pure transport. A second transport 707 was lost in a training accident off East Sale Vic in October 1991.

The four tankers are aircraft originally built for Qantas as 707-338Cs. After being sold by Qantas they served with other operators before returning to the Australia for the RAAF in 1979 (two aircraft) and 1983. They were originally employed by the RAAF as transports but a request for tender for the conversion of the aircraft to tanker/transports was issued in 1987.

The $15m contract was won by Israeli Aircraft Industries who, in association with its Australian partner, Hawker de Havilland, carried out the conversions which involve the installation of wing tip pods for the US Navy style 'hose and drogue' refuelling system compatible with the Hornet. The first converted aircraft was handed back to the RAAF in February 1990 and all the 707s are operated by No 33 Squadron based at Richmond.

The tankers are an essential element of Australia's tactical fighter force, the oft used phrase 'force multiplier' accurately describing their effect. Unfortunately, the 707 tankers cannot be used with the RAAF's F-111 bombers, as they are equipped with the USAF 'flying boom' refuelling system. The F-111s therefore have to rely on visiting USAF tankers for their aerial refuelling needs.

An RAAF press release gives an example of the use the tankers in conjunction with Hornets, albeit slightly poetically and very much in 'PR speak': "First Non Stop To Butterworth for F/A-18s – Six 3 Squadron F/A-18s make their final approach into Butterworth, emerging out of the hazy atmosphere, touching down as though they do it every day. Just another deployment to Malaysia?

"Not quite. Pilots from the RAAF's three fighter squadrons – 3, 75 and 77 – are accustomed to long flights across Australia's deserts and over the Timor Sea to the Malaysian Peninsular. But normally the trip is broken with ground stops for refuelling. On 18 September however, when those six 3 Squadron pilots strapped into their cockpits, they knew they weren't going to touch mother earth again until they had reached Butterworth.

"It was to be the first time RAAF F/A-18s attempted the non stop, 7,000km flight, using a network off 33 Squadron Boeing 707 tankers to achieve their goal. With careful planning over a number of weeks, and the inevitable last minute changes, two operational elements of the RAAF normally carrying out totally different roles, meshed successfully in the air, in what has been hailed as a milestone for both the Tactical Fighter Group and the Air Lift Group.

"For the non stop flight, according to 33 Squadron mission co-ordinator Flt Lt Craig Fulton, one 707 launched from Williamtown and another from Darwin to offload a combined total of 35,000lb (15,900kg) of fuel for each fighter over five separate mid air 'hook ups'.

"Craig said the project was the most ambitious undertaken so far by the two groups – and was subject to last minute

One of the best assets the RAAF has acquired in recent years is Boeing 707 tankers, a 'force multiplier' which allows the Hornet's considerable potential to be more fully realised.

Assorted 3 Squadron Hornets at Darwin in March 1988 (top to bottom): A21-18, A21-13, A21-108 and A21-16. (Bill Lines)

changes as is normally expected with any mission. One element over which the planners had absolutely no control was the weather. Mother Nature played one of her usual tricks, breathing headwinds of almost 150 knots over northern Australia.

"'We had planned for an 85 per cent chance of headwind component – around 60 or 70 knots for that first leg', said Craig. He said the strong winds meant precious fuel supplies were being burned up more quickly, resulting in the tanker giving away more to the F/A-18s than had originally been planned. With their own stocks running low, the tanker crew diverted to Mt Isa to refill before continuing.

"The amount of fuel delivered to the 'receivers' was significant, according to Craig. 'The F/A-18s took off with 17,000lb (7,718kg), but by themselves they would have only made it to somewhere between Tindal and Mt Isa. You might say we more than trebled their range'.

"... despite the usual banter between 'trashies' and 'knucks' (transport and fighter pilots) and the tendency to get a little scathing (albeit with good humour) about who does what and how well, the touchdown of the 707 at Butterworth saw an unprecedented gesture of appreciation from some grateful 3 Squadron pilots after their nine-and-a-half hours in the cockpit. The tanker crew was presented with some delightfully soothing, cold liquid refreshments ..."

First Test

Australia's need for a tactical fighter which can fulfil a number of roles is met by the versatile Hornet, and the day to day operations of the RAAF's three fighter squadrons reflect this. All peacetime flying is in effect training, and the RAAF's Hornets and their crews practice skills which hopefully won't be needed either by themselves or in conjunction with other RAAF and foreign aircraft.

Exercises are a major part of any military training programme, the larger ones usually involving land, sea and air elements from several nations and lasting some weeks.

The first major exercise in which RAAF Hornets took part was 'Pitch Black' in 1987, an air defence exercise based in Darwin where RAAF fighters had to fight off attacking forces comprising RAAF F-111s and USAF F-4 Phantoms and B-52 bombers, all supported by various other RAAF, USAF and RAN aircraft including electronic warfare aircraft which attempted to 'jam' the fighters' flow of information. The Hornets came from No 3 Squadron and they were supported by Mirages from Nos 75 and 77 Squadrons. This was the last major exercise in which RAAF Mirages took part.

More 3 Squadron Hornets (top to bottom): A21-12 and A21-9, both at Darwin in July 1988; A21-36 at Richmond in October 1988 for the Bicentennial Air Show and A21-17 (with LEX strakes fitted) at Darwin in July 1990. (Bill Lines)

Pitch Black was a good initial test for the new Hornets and involved combat air patrols (CAPs), constant pressure, and a capable foe. An important part of the assessment of the Hornet's performance was how it, its pilots and the fighter controllers on the ground handled the enormous quantity of information available to them by quickly assessing it, making the correct tactical decisions and then scoring a 'kill' or at least forcing the attacking enemy to think again.

After this first major test of the Hornet in the field – when 3 Squadron was still below full strength in both aircraft and pilots – the verdict on the aircraft was very positive. Although that particular Exercise Pitch Black involved interception rather than dogfighting aerial combat, the RAAF's Hornets have proved to be more than a match for most types in contests against USAF F-4 Phantoms, F-15 Eagles and other types. The Eagle is claimed to be the best dogfighter in the world and the Hornet has proved to be capable of holding its own in one-on-one fights. It has been suggested that the Hornet is superior to the Eagle at lower speeds due to its turning ability, a function of the large manoeuvring flaps and the powerful and responsive F404 engines.

Reliability

While the performance of any aircraft in the air is what ultimately counts, its performance on the ground – its maintenance and reliability characteristics – are equally important. If an aircraft can't fly because of some unserviceability or another it is of no use to a military commander.

The Hornet has always exhibited excellent characteristics in this area, although it is true to say that in the early days of RAAF service the target ratio of maintenance hours per flight hour was not met due to factors such as a shortage of spares and trained maintenance personnel. That has since improved to the expected standard.

The Hornet's design makes it inherently more reliable than earlier generation fighters due to its built in test equipment (BITE), a self checking system which enables faulty components to be quickly identified and replaced or repaired. The result is an aircraft which spends less time on the ground and more time in the air doing the job for which it was designed.

US Navy figures compiled after several years' service tend to confirm the Hornet is a reliable aeroplane. Mean flight hours between failure (of avionics, systems etc) for the Hornet was 3.6 (2.2 in its early days of service) compared with only 0.7 for the A-7E Corsair, 0.6 for the A-6E Intruder, 0.9 for the F-4 Phantom and 0.5 for the F-14A Tomcat.

77 Squadron Hornets photographed at Darwin at various times (top to bottom): A21-1, A21-114 and A21-35 in July 1989 plus A21-4 in July 1988. (Bill Lines)

The Hornet's maintenance hours required per flight hour statistic is equally impressive when compared with other types: Hornet 16, Intruder 52, Corsair 42, Tomcat 55 and Phantom 42. These figures will of course vary depending on the time of the survey and the prevailing conditions, but even in its earliest days of US Navy service the Hornet was generating just 24 maintenance hours per flight hour, still substantially better than the other aircraft.

The Hornet has also proved to be a safe aircraft. By late 1989 US Navy figures showed the loss rate to be 4.5 per 100,000 flight hours, again a substantial improvement over other aircraft: Tomcat 10, Intruder 13, Phantom 19, Corsair 22 and A-4 Skyhawk 29.

At the time of writing (February 1993) the RAAF had lost four Hornets in accidents after a period of service approaching eight years. By comparison, a dozen Mirages had been lost after the same period of RAAF service.

Some Other Activities

It's only possible to give an overview of RAAF Hornet activities in this publication, but the following gives some idea of the variety of duties performed by the aircraft as part of their day to day lives.

In April 1992 an exercise was held in which US Navy F-14 Tomcats and A-6 Intruders from the visiting aircraft carrier USS *Independence* launched an 'attack' on RAAF Williamtown as part of war games being carried out by the Australian and US military during Exercise Coral Sea.

The exercise involved Williamtown based Hornets from No 3 Squadron defending their base against the attacking American aircraft and also retaliating with an attack on the carrier. The Williamtown Air Defence Radar Unit was of course involved, this type of exercise being a vital part of keeping up the required level of co-ordination and proficiency necessary if a real attack was ever to occur.

An interesting exercise involving four 77 Squadron Hornets, a Boeing 707 tanker and Navy mine clearance divers from HMAS *Waterhen* (a Sydney shore establishment) was carried out during September 1992.

The Hornets flew from Williamtown to the Halifax Bay Training area near Townsville, Queensland, to drop between them eight inert Mk.36 destructor mines. It was the first time aerial mines have been despatched from a jet aircraft in Australia – a job normally performed by maritime patrol aircraft – the idea being to demonstrate the Hornet's capability in this area of operation.

The 1,800nm (3,330km) mission was flown non stop and involved four hours of flying and an aerial refuelling in each direction between the Queensland towns of

More 77 Squadron Hornets at Darwin (top to bottom): A21-5 in July 1988 plus A21-53, A21-51 and A21-33 in July 1990. (Bill Lines)

Roma and Emerald. The Navy mine clearance divers were used to remove the mines from the water and to assess the accuracy of the drop.

In June 1992, Rockhampton Qld airfield took on the atmosphere of an RAAF base when five 77 Squadron Hornets joined six 76 Squadron Macchis and three Winjeel forward air control aircraft from the same unit for a combined exercise called 'Tasman Link'.

The Hornets flew maritime air defence sorties and the Macchis were used for close air support in the exercise, which was held in and around the Shoalwater Bay Training Area.

In October and November 1992, Exercise Aces North, a major air combat exercise held in the Northern Territory brought some 1,500 visiting military personnel to the RAAF bases at Tindal and Darwin. The exercise involved 70 aircraft from the RAAF, USAF and US Marine Corps and was designed to provide a realistic scenario involving the defence of the Delamare Weapons Range near Tindal against enemy forces which included RAAF F-111s and Hornets plus USAF and USMC Hornets, EA-6B Prowlers and F-15 Eagles. Ground forces were also involved in the exercise.

Defending the range were RAAF Hornets from 75 Squadron at Tindal working in conjunction with a pair of air defence radars deployed in the field and Tindal's air traffic control radar.

Political Interference

A shortage of pilots in the late 1980s/early 1990s when the RAAF was steadily losing aircrews to the airlines has had some effect on RAAF Hornet operations (for a time there were more Hornets available than pilots), as have budget cuts.

In 1990 the Australian government announced that Hornet (and F-111) flying hours had to be reduced by 7.5 per cent, a situation which could have meant reduced long term operational effectiveness for the front line squadrons. The figures are not publicly available, but it is estimated by experts that about 200-250 hours per annum is a typical Hornet utilisation figure.

Given the variety of tasks the aircraft can perform and indeed needs to perform, this is a relatively low figure, and the cuts applied only to operational squadrons, not training or conversion units such as 2 OCU. Prior to the announcement of the cuts, the RAAF's Hornet operational and training schedule was already very tight but the move forced new efficiencies to be found and incorporated into the system without compromising operational effectiveness.

75 Squadron Hornets at Darwin (top to bottom): A21-26, A21-29 and A21-116 in May 1989 plus A21-29 again, this time sporting its traditional top hat and cane motif in enlarged form. (Bill Lines)

Accidents

As mentioned above, after some eight years of service the RAAF had lost only four of its 75 Hornets, two examples each of single seat F/A-18As and two seat F/A-18Bs.

The first loss occurred in November 1987 when 2 OCU's A21-104 hit a hill on Great Palm Island near Townsville Qld during a night radar navigation and bombing sortie. The aircraft was destroyed and its pilot killed, the accident occurring during the radar bombing pattern phase of the sortie.

The investigation into the accident revealed pilot error to be the cause, an error in the management of the Hornet's navigation system resulting in a non standard entry into the bombing pattern, causing the downwind leg to be considerably closer to the island than had been briefed. Although the aircraft turned towards the target near the planned distance of 10nm, its incorrect placement caused it to track over Great Palm Island and into the mountain.

The next loss occurred in August 1990, A21-42 and A21-29 from 75 Squadron at Tindal colliding during air combat manoeuvring exercises about 30 kilometres from the base.

Hornet A21-29 recovered to Tindal with part of its outer wing missing, while A21-42 crashed with the loss of Wng Cdr Ross Fox, the squadron's commanding officer and one of the RAAF's leading and most accomplished fighter pilots.

Very early in the Hornet's RAAF career – in February 1986 – two aircraft from 2 OCU (A21-102 and 107) had a very close shave when they collided during a 2 versus 1 intercept exercise. Failure to maintain visual contact was given as the primary reason for the incident which ended happily with both aircraft landing safely albeit with damage.

The loss of A21-41 (another 75 Squadron Hornet) in June 1991 has provided one of the great mysteries of air safety in recent times.

Flg Off Cameron Conroy's aircraft disappeared after he apparently became incapacitated and unable to control his aircraft during a routine navigation sortie in the Northern Territory.

The pilot of a second Hornet escorted Conroy's aircraft and issued a 'Mayday' call until he was forced to return to Tindal due to low fuel. Conroy was seen to be apparently slumped in the seat of his aircraft, which continued to fly on, straight and level. A third Hornet was despatched to join A21-41 but it was unable to catch up before the stricken aircraft disappeared off radar over the Gulf of Carpentaria.

A massive search was carried out covering a large area over both land and sea but

Top to bottom: Hornets A21-34 and A21-6 of 75 Squadron at Darwin in September 1988 and Williamtown during 1990, respectively; 2 OCU's A21-114 at Richmond in July 1990; A21-101 in ARDU markings at Edinburgh in November 1991. (Bill Lines/Bruce Chapman)

was unsuccessful. The RAAF used the Hornet simulator at Tindal in an attempt to recreate the aircraft's path after it ran out of fuel, but despite this, the size of the search and the fact that people in the Weipa area reported seeing an aircraft condensation trail north of the town on the day of the mishap, the wreckage and pilot's body have never been found, nor has the cause of the incapacitation been positively identified.

The fourth and most recent RAAF Hornet loss (to early 1993) was that of 77 Squadron's A21-106 at Cape Clinton, North of Rockhampton Qld in May 1992. Both crew members – Flg Off Robert Jepsen and his passenger, civilian defence scientist Michael Howlett – lost their lives.

A21-106 was operating out of Townsville on a training exercise in conjunction with other aircraft in the Shoalwater Bay Military Training Area when the accident occurred.

Mods and Upgrades

Like all aircraft which will have a lengthy period of service before being replaced, the RAAF's Hornets have and will be subject to upgrades to increase their operational effectiveness and to cure problems.

These include structural modifications to increase airframe life and installing provision for advanced avionics. Shortly after the last new Australian built Hornet was handed over to the RAAF in May 1990, work began on modifying earlier aircraft at the factory from which they were first rolled out at Avalon, Victoria.

By early 1993 the programme was well on the way to completion, the idea being to bring the aircraft up to the standard of the F/A-18C and D model Hornets. This not only makes the RAAF's Hornets more capable, but also ensures operational and logistical compatibility with US and other nations' Hornets.

The first aircraft to be modified was A21-102, work on which was completed in September 1990. Upon arrival at Avalon, the aircraft are stripped of their wings, horizontal tail surfaces, engines and cockpit before being reassembled and reflown to their new standard. The work takes eight to 12 weeks to complete.

The modification programme being undertaken by ASTA at Avalon comes under the heading Engineering Change Proposals (ECP) and include the following, all of which are designed to ensure that Australia's Hornets are able to accept the latest internal modifications as they become available. They would, for example, make the aircraft capable of using the new

Hornets A21-1, -107 and -101 of No 2 Operational Conversion Unit. (Defence PR)

When we grow up ... a Hornet makes an appearance at an airshow, taxying in among some much less potent distant cousins.

AIM-120 advanced medium range air-to-air missile (AMRAAM) later in the decade:
ECP 176C2: provision for laser target designator ranger and adds an LTD/R arm switch to the sensor control panel, along with associated wiring. This will be used in association with the Ford Aerospace FLIR which entered service in 1991.
ECP 228R1: A retrofit of FMS (flight management system) advanced systems provision which allow the fitting of new mission computers (XN-6), a new armament control processor plus additional or modified weapons interface wiring. Also provided for are laser blanking options and the incorporation of a new stick adaptor plus a modified data storage set mounting with associated wiring provisions.
ECP 228R1C1: An FMS advanced systems provision which essentially provides the capability to load electronic warfare equipment via a nosewheel well connector.
ECP 287: The incorporation of various advanced systems including new mission computers, new armament control processor set, modified digital display indicators, improved stick grip, new data storage set, modified pylons, new adaptor cables and new data transfer equipment.
ECP 272R1: Transmitter consolidation with an upgraded microprocessor providing full frequency agility with a redesigned grid modulator for improved reliability. The receiver/exciter also has a redesigned digital mode for enhanced reliability while the radar target data processor will have modified timing and control SRAs. The computer will also have its disc memory replaced with EEPROM memory for better reliability and future growth.

The modifications which have taken place so far revolve around the installation of the Ford Aerospace AN/AAS-38 FLIR (with laser target designator capability available in 1993) and fitting a Northrop AN/ALQ-162 receiver/jammer. Upgrading the Hughes AN/APG-65 radar to AN/APG-73 standard will come later. The new radar was first flown in a US Navy Hornet in April 1992 and will become available during 1994.

In addition, at some stage the Hornet's engines will be upgraded from the original F404-400 standard to the more powerful -402 as is fitted to the F/A-18C and D.

The Hornet's 'office' – MFDs, HUD, HOTAS – all the mod cons.

Fatigue Testing

The question of fatigue life is an important one, so much so that in 1992 the RAAF and the Canadian Forces began a joint fatigue testing programme on the F/A-18A. Called the International Follow On Structural Test programme (IFOSTP), the aim of the project is to determine and possibly extend the service lives of the Australian and Canadian 'A' model Hornet fleets. Canada's Hornets are currently scheduled to retire in 2003 and Australia's from about 2010.

Fatigue testing conducted by McDonnell Douglas was based on the US Navy's expected usage and operational conditions, resulting in an estimated airframe life of 6,000 hours. But the nature of RAAF and CAF operations has significantly reduced this. IFOSTP aims to validate the airframe by applying RAAF/CAF operating factors and by determining possible structural failure points hopes to be able to restore the 6,000 life and possibly extend it by further testing.

In Canada, Bombardier and the National Research Council are undertaking testing of the fuselage centre section and wings with a view to establishing a safe structural life for critical items such as the wing attachment bulkheads. A test rig applies a combination of aerodynamic and inertia loads in a programme which simulates 12,000 hours of service operation.

Australia's testing of the aft fuselage and empennage simultaneously applies dynamic buffet loads and manoeuvre static loads to duplicate the environment to which the airframe is exposed in service. The testing is carried out by the Defence Science and Technology Organisation's (DSTO) Aeronautical Research Laboratory (ARL). Again, 12,000 hours will be simulated. In both cases, inspection at 12,000 hours will be followed by continuation of the programmes until an uneconomic repair life is established.

As discussed in the opening Hornet chapter, one of the aircraft's main problems has been vortices from the wing leading edges extensions (LEX) impacting the horizontal and vertical tail surfaces and creating large buffeting loads and therefore cracking. A modification was devised – fitting a fence on each LEX – and although this reduced the problem it did not solve it.

Canada and Australia have had previous discussions about the Hornet. In 1988 the Canadian government approached Australia about purchasing 25 locally built Hornets. Like Australia, Canada operates A and B models and as US production had by then switched to the upgraded C and D variants, Australia was the only production source of the earlier models.

At the time, Australia had hopes of selling locally built Hornets to other air forces in the region, notably New Zealand and Singapore. A Canadian purchase would not only

Hornet A21-12 turns in for a low and slow pass over Adelaide's Formula One Grand Prix street circuit during a pre race display.

extend the Hornet manufacturing programme by 18 months but it would also give the RAAF the option of purchasing some C and D models, incorporation of which into the Australian production line was said to be not difficult. As it turned out, no Australian Hornets were exported to Canada or any other country.

The RAAF's Hornets have been grounded once during their period of service. In October 1990 all aircraft (and those serving with other nations) had to be checked following the discovery of a wiring problem in a US Navy aircraft.

The problem stemmed from electrical cables carrying the aircraft's 115 volt power supply from the port generator into the main power system sagging down in flight and touching a fibreglass duct. Left to its own devices, the loom in question would eventually wear down, expose the wires, arc and cause a fire. Inspection of the RAAF's fleet revealed that all but a few of the Hornets were in fact suffering from the problem to some extent. In most cases repairs took only three or four hours to complete and involved strapping the wiring up and away from the ducting, although two aircraft in which the wiring had corroded took longer to modify.

An interesting shot of A21-103 as it turns onto finals with everything down. (RAAF Historical)

The Future

By 1993 the history of the Hornet in RAAF service was stili very young – only about one-third of the aircraft's planned service life had been completed and the type will serve well into the 21st century.

There has been speculation that more Hornets (the very capable F/A-18C and D 'Night Attack' models) will be acquired at some stage to replace the RAAF's General Dynamics F-111s in the strike role, although the surprise announcement in late 1992 that 18 additional second hand F-111s would be purchased from USAF stocks as attrition replacements for the existing fleet at a bargain basement price appears to nullify that idea. On the other hand, at the time of writing, some were expressing doubts that the F-111 deal would be consumated.

Either way, the McDonnell Douglas F/A-18 Hornet will continue to be one off the most important assets in Australia's defence armoury for many years to come.

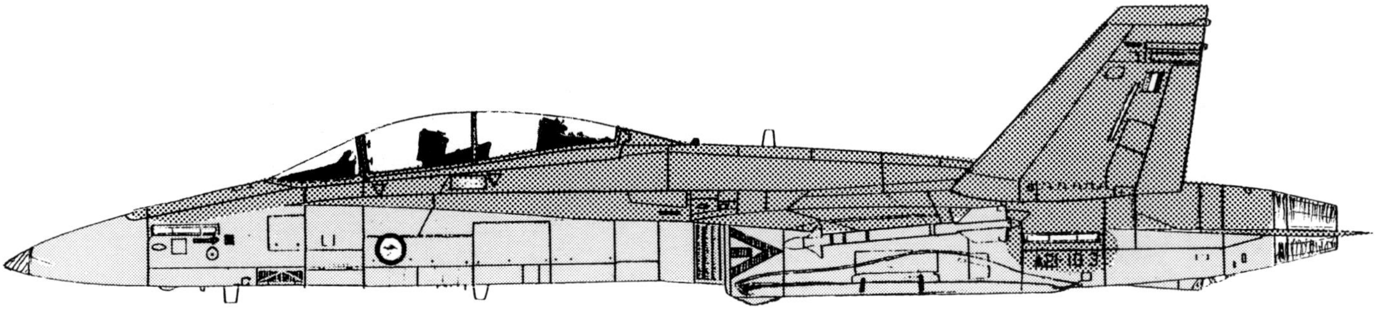

F/A-18B Hornet A21-103, the first Australian assembled Hornet, as it appeared at its debut, 1984
Standard camouflage scheme of blue grey upper surface with light grey undersurface. Black serial number. Tan nose tip.

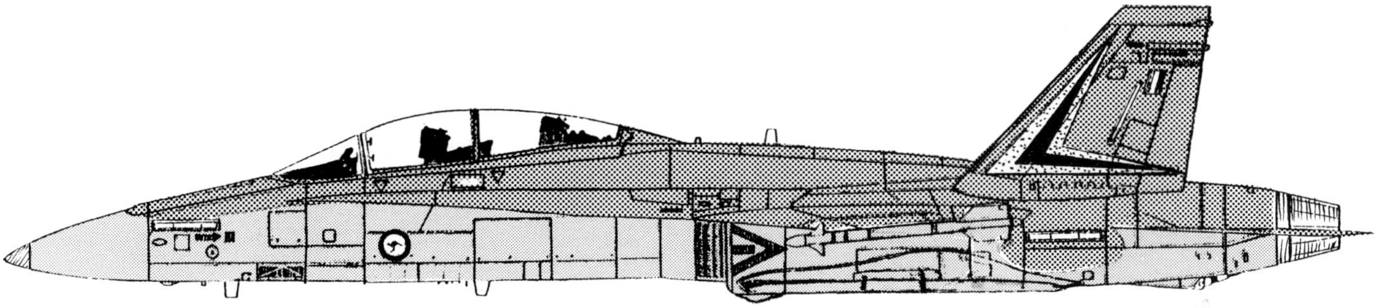

F/A-18B Hornet, A21-101, in delivery marking from USA, 1985
Standard scheme. Yellow and black fin flash. Note the absence of any serial.

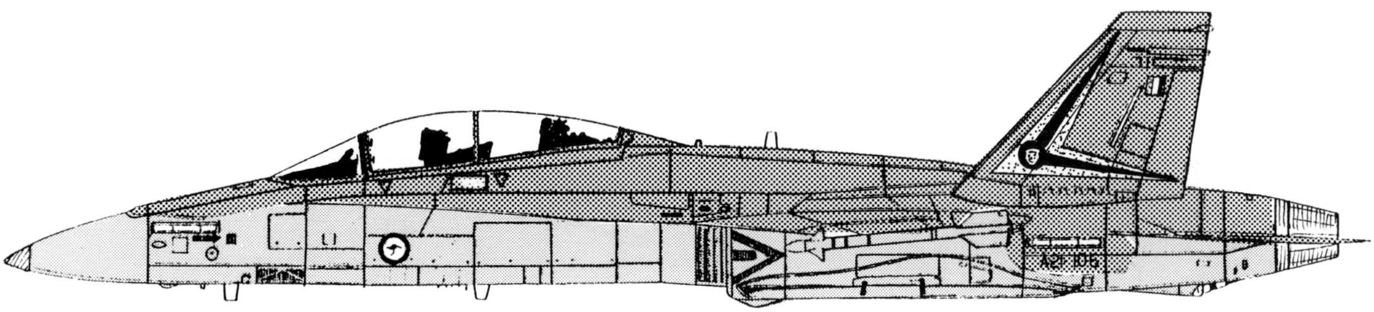

F/A-18B Hornet, A21-105 of No 2 OCU, 1986
Standard scheme. Yellow and black fin flash. Squadron emblem on black disc at base of flash.

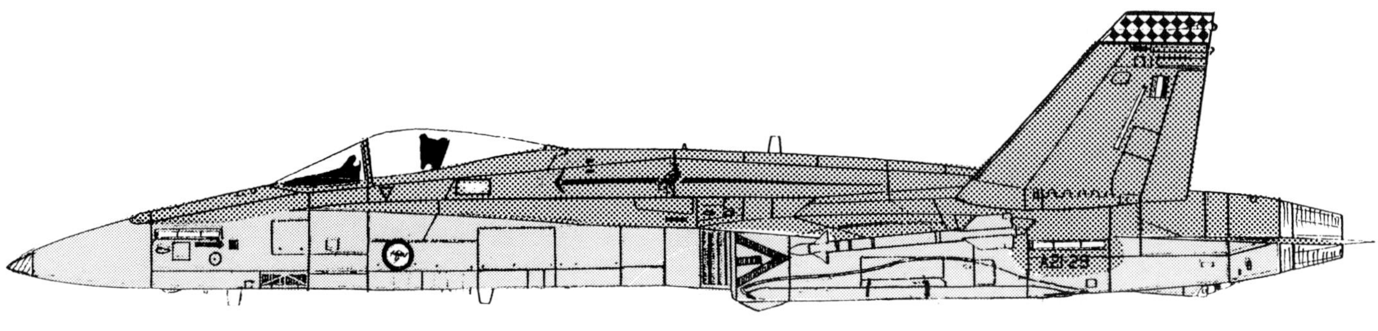

F/A-18A Hornet A21-29 of No 75 Sqn, late 1980s
Standard scheme. Black fuselage flash with black and white magpie. Black and white diamonds, with black borders, on fin tip.

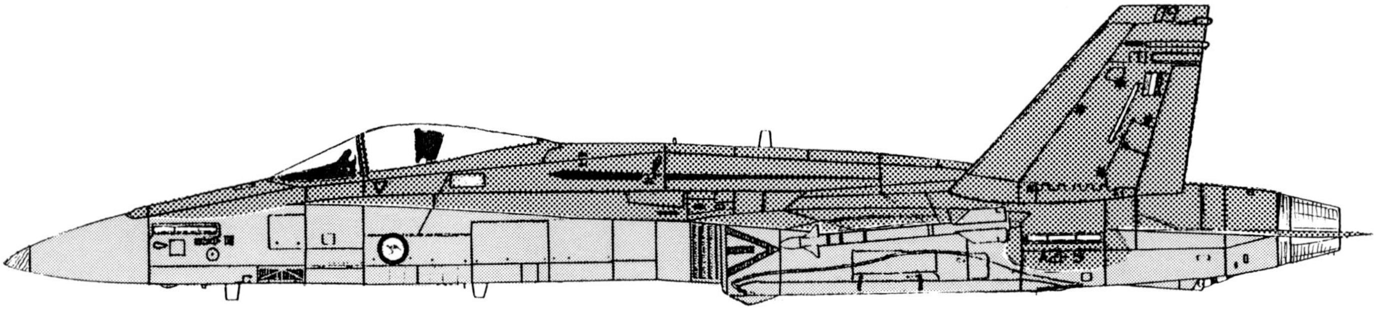

F/A-18A Hornet A21-9 of No 3 Sqn, 1987
Standard scheme. Black flash and fleur-de-lis on fuselage. Black stars on fin representing the Southern Cross Constellation. Serial is repeated on fin tip (note the 0 prefix).

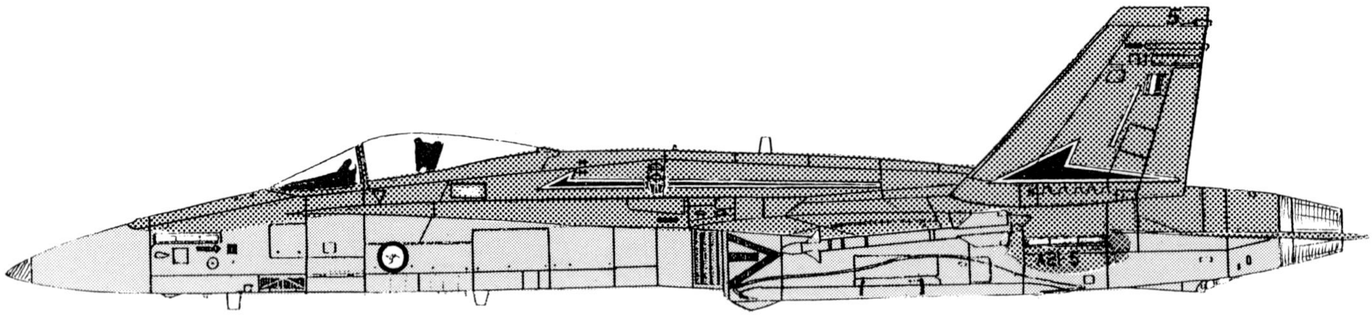

F/A-18A Hornet A21-5 of No 77 Sqn, late 1980s
Standard scheme. Dark green squadron markings on fuselage and fin, all with white border and thin green outline. Serial is repeated on fin tip.

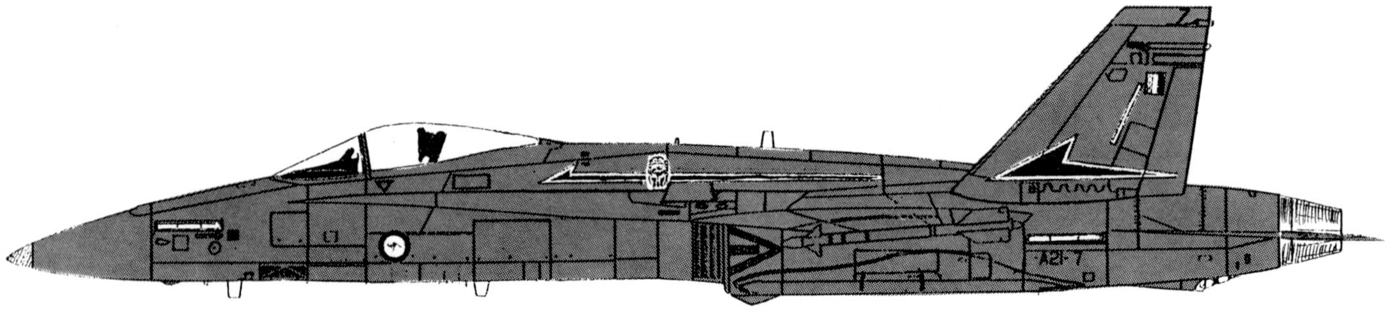

F/A-18A Hornet A21-7 of No 77 Sqn, 1990
Experimental camouflage scheme of overall dark grey. Standard squadron markings. Serial is repeated on fin tip.

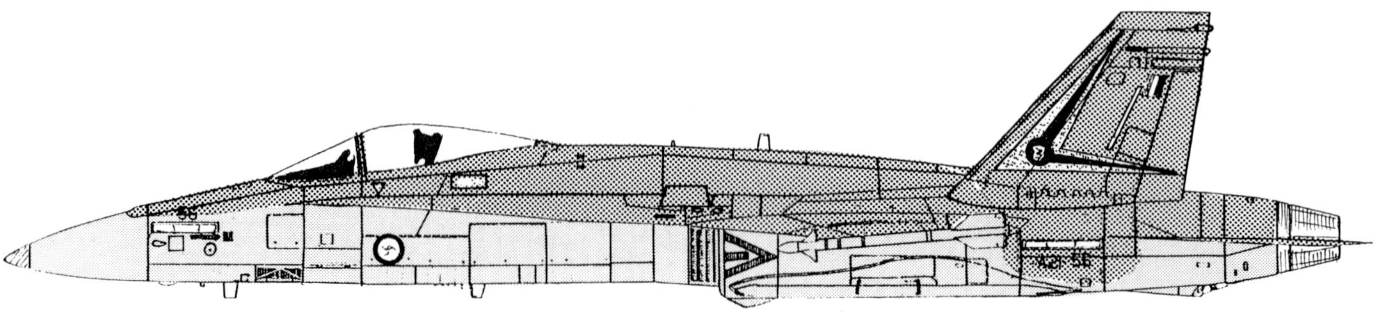

F/A-18A Hornet A21-56 of No 2 OCU, 1992
Standard scheme. Yellow and black fin flash with squadron emblem on a black disc at base of flash. Serial number (56) is repeated on nose above formation light.

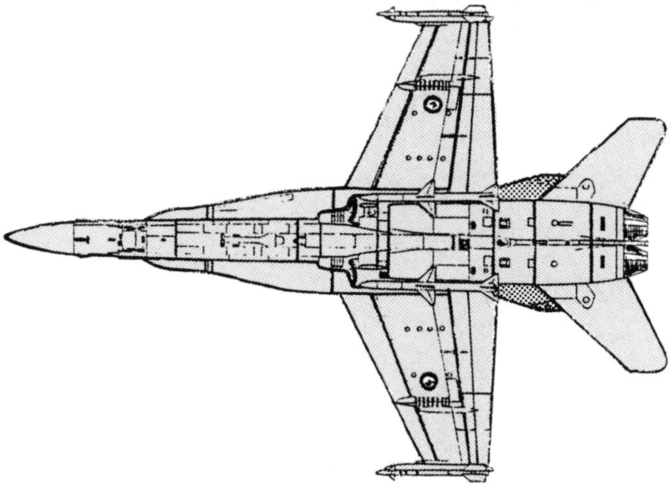

Undersurface, showing location of roundels. No roundels are applied on the blue grey upper surface.

Marking applied to fins to No 2 OCU Hornets. Tiger has yellow face, black details, white eyes and fangs and red mouth.

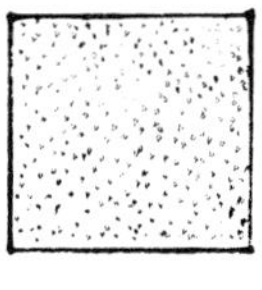

Yellow

Dark Grey (FS 36118)

Light Grey (FS 36375)

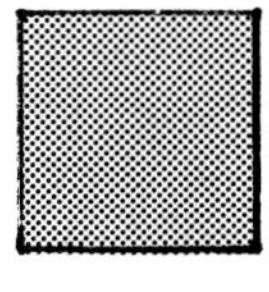

Blue Grey (FS 35237)

RAAF McDONNELL DOUGLAS F/A-18 HORNETS

Abbreviations: C/No – constructor's number; Sqdn – squadron; OCU – Operational Conversion Unit; deliv – delivery; MD – McDonnell Douglas; ARDU – Aircraft Research and Development Unit; ff – first flight.

Notes: The 'in service' note in the Disposal/Remarks column reflects the individual aircraft's status in 1992 and may well have changed by the time this is published. It nevertheless gives a general idea of how the Hornets are deployed within the RAAF.

The 'MD No' column (McDonnell Douglas number) refers to the manufacturer's F-18 general production run aircraft number.

RAAF No	Model	C/No	MD No	Deliv	Disposal/Remarks
A21-1	F/A-18A	AF-1	228	11/85	in service 77 Squadron
A21-2	F/A-18A	AF-2	245	02/86	in service 77 Squadron
A21-3	F/A-18A	AF-3	250	03/86	in service 77 Squadron
A21-4	F/A-18A	AF-4	263	06/86	in service 77 Squadron
A21-5	F/A-18A	AF-5	271	06/86	in service 77 Squadron
A21-6	F/A-18A	AF-6	287	07/86	in service 3 Squadron
A21-7	F/A-18A	AF-7	292	06/86	in service 77 Squadron
A21-8	F/A-18A	AF-8	306	08/86	in service 3 Squadron
A21-9	F/A-18A	AF-9	316	08/86	in service 3 Squadron
A21-10	F/A-18A	AF-10	326	10/86	in service 3 Squadron
A21-11	F/A-18A	AF-11	334	10/86	in service 3 Squadron
A21-12	F/A-18A	AF-12	339	10/86	in service 3 Squadron
A21-13	F/A-18A	AF-13	345	07/87	in service 3 Squadron
A21-14	F/A-18A	AF-14	353	12/86	in service 2 OCU
A21-15	F/A-18A	AF-15	361	01/87	in service 77 Squadron
A21-16	F/A-18A	AF-16	368	04/87	in service
A21-17	F/A-18B	AF-17	376	05/87	in service 3 Squadron
A21-18	F/A-18A	AF-18	382	04/87	in service 3 Squadron
A21-19	F/A-18A	AF-19	454	11/87	in service 77 Squadron
A21-20	F/A-18A	AF-20	462	12/87	in service 77 Squadron
A21-21	F/A-18A	AF-21	470	01/88	in service 2 OCU
A21-22	F/A-18A	AF-22	476	01/88	in service 77 Squadron
A21-23	F/A-18A	AF-23	483	02/88	in service 2 OCU
A21-24	F/A-18A	AF-24	489	02/88	in service 75 Squadron
A21-25	F/A-18A	AF-25	496	05/88	in service 2 OCU
A21-26	F/A-18A	AF-26	506	03/88	in service 75 Squadron
A21-27	F/A-18A	AF-27	514	05/88	in service 75 Squadron
A21-28	F/A-18A	AF-28	521	05/88	in service 2 OCU
A21-29	F/A-18A	AF-29	527	04/88	in service 75 Squadron
A21-30	F/A-18A	AF-30	535	05/88	in service 3 Squadron
A21-31	F/A-18A	AF-31	552	06/88	in service 75 Squadron
A21-32	F/A-18A	AF-32	560	08/88	in service ARDU
A21-33	F/A-18A	AF-33	569	06/88	in service 75 Squadron
A21-34	F/A-18A	AF-34	577	07/88	in service 75 Squadron
A21-35	F/A-18A	AF-35	588	07/88	in service 77 Squadron
A21-36	F/A-18A	AF-36	595	07/88	in service 2 OCU

RAAF No	Model	C/No	MD No	Deliv	Disposal/Remarks
A21-37	F/A-18A	AF-37	640	10/88	in service
A21-38	F/A-18A	AF-38	651	10/88	in service 3 Squadron
A21-39	F/A-18A	AF-39	657	11/88	in service 2 OCU
A21-40	F/A-18A	AF-40	665	12/88	in service 2 OCU
A21-41	F/A-18A	AF-41	674	12/88	lost without trace Gulf of Carpenteria area 06/91
A21-42	F/A-18A	AF-42	680	07/89	collided with A21-29 near Tindal NT 08/90 and crashed
A21-43	F/A-18A	AF-43	689	02/89	in service 75 Squadron
A21-44	F/A-18A	AF-44	707	06/89	in service 75 Squadron
A21-45	F/A-18A	AF-45	722	07/89	in service 77 Squadron
A21-46	F/A-18A	AF-46	729	06/89	in service 77 Squadron
A21-47	F/A-18A	AF-47	741	07/89	in service 75 Squadron
A21-48	F/A-18A	AF-48	750	08/89	in service 77 Squadron
A21-49	F/A-18A	AF-49	757	09/89	in service 75 Squadron
A21-50	F/A-18A	AF-50		09/89	in service 77 Squadron
A21-51	F/A-18A	AF-51		10/89	in service 77 Squadron
A21-52	F/A-18A	AF-52		11/89	in service 3 Squadron
A21-53	F/A-18A	AF-53		12/89	in service 77 Squadron
A21-54	F/A-18A	AF-54		01/90	in service 3 Squadron
A21-55	F/A-18A	AF-55		03/90	in service 3 Squadron
A21-56	F/A-18A	AF-56		04/90	in service 2 OCU
A21-57	F/A-18A	AF-57		05/90	in service 3 Squadron
A21-101	F/A-18B	ATF-1	213	03/85	first RAAF Hornet, ff (in USA) 13/08/84, in service ARDU
A21-102	F/A-18B	ATF-2	214	03/85	US assembled, in service 2 OCU
A21-103	F/A-18B	ATF-3	215	04/85	first Australian assembled Hornet, ff 26/02/85, in service 2 OCU
A21-104	F/A-18B	ATF-4	216	07/85	crashed Great Palm Island Qld 11/87
A21-105	F/A-18B	ATF-5	217	10/85	in service 2 OCU
A21-106	F/A-18B	ATF-6	224	09/85	crashed near Cape Clinton Qld 05/92
A21-107	F/A-18B	ATF-7	235	12/85	in service 3 Squadron
A21-108	F/A-18B	ATF-8	392	06/87	in service 3 Squadron
A21-109	F/A-18B	ATF-9	398	09/87	in service
A21-110	F/A-18B	ATF-10	413	09/87	in service 2 OCU
A21-111	F/A-18B	ATF-11	423	11/87	in service 2 OCU
A21-112	F/A-18B	ATF-12	427	09/87	in service 2 OCU
A21-113	F/A-18B	ATF-13	436	09/87	in service 2 OCU
A21-114	F/A-18B	ATF-14	443	11/87	in service 77 Squadron
A21-115	F/A-18B	ATF-15	604	09/88	in service 2 OCU
A21-116	F/A-18B	ATF-16	614	08/88	in service
A21-117	F/A-18B	ATF-17	623	09/88	in service 2 OCU
A21-118	F/A-18B	ATF-18	632	12/88	in service

McDONNELL DOUGLAS
A-4 SKYHAWK

A fraction of a second before touchdown, a Royal Australian Navy A-4G Skyhawk catches the wire on HMAS Melbourne. *(RAN)*

McDONNELL DOUGLAS A-4 SKYHAWK

"A small aeroplane designed to do the job of a medium sized aeroplane" is one writer's description of the Douglas A-4 Skyhawk, designed in the early 1950s as a naval attack aircraft intended to reverse the trend towards ever larger and heavier aircraft.

'Heinemann's Hot Rod', 'The Scooter' and 'Tinker Toy Bomber' were some of the nicknames applied to the (relatively) diminutive Skyhawk which first flew in 1954 and must have achieved many of its designer's aims as production ended 2,960 examples and a quarter of a century later; the longest production run of any US combat aircraft.

In between, the Skyhawk successfully fought in several wars – Vietnam, the Middle East and the Falklands – and is still in front line service with several nations. For an aircraft which flew in the face of conventional wisdom at the time of its conception, that is a remarkable enough record; that the Skyhawk's airframe and concept is still considered worthy of being modified and updated to create a combat aircraft with a useful part to play in the last years of the 20th century adds to its reputation.

The Skyhawk's overriding design philosophy was one of compact size, light weight, low cost and general simplicity and here it succeeded admirably. Its maximum weight was initially only half that considered necessary by the US Navy operational requirement which inspired it and in service it was capable of generating a high sortie rate with modest maintenance requirements supporting it. It also proved capable of absorbing considerable battle damage and surviving.

But designer Ed Heinemann's determination to keep everything to a minimum brought with it compromises, an inevitable and immovable law which applies to any aircraft. It was very much a single role aircraft (although later updates have enabled it to operate in a limited defensive role as well as that of attack), its aerodynamics have been described as "unremarkable" (turn performance suffers as does speed and rate of climb) while the small airframe resulted in a general lack of space and by American standards, a small cockpit.

This is all very well, but the fact remains the Skyhawk *was* successful, *did* achieve its designer's aims, influenced other designers and has the production run and operational history behind it as proof.

The Skyhawk's significance to Australia lies in the fact that it was the final combat aircraft to be operated from an aircraft carrier by the Royal Australian Navy's Fleet Air Arm. Twenty Skyhawks were delivered to the RAN (ten each in 1967 and 1971) replacing de Havilland Sea Venoms, these aircraft operating from the flagship HMAS *Melbourne* until 1982 when the carrier was decommissioned. The ten surviving Skyhawks were sold to the Royal New Zealand Air Force in 1984 and subsequently upgraded.

Heinemman's Hot Rod

The Skyhawk came about as the result of a US Navy requirement for a replacement for the successful Douglas AD (later A-1) Skyraider series of single piston engined close support aircraft, some 3,180 of which were built between 1945 and 1957 in single and two seat versions for both shipboard and land based operations by the Navy and

The prototype Douglas XA4D-1 Skyhawk first flew on 22 June 1954, the ancestor of 2,959 further examples and a production life spanning 25 years.

The Skyhawk came about as the result of the US Navy's search for a successor to the Douglas A-1 Skyraider, 3,180 of which were built between 1945 and 1957. (USN)

the US Air Force. The Skyraider, of course, had continued Douglas' wartime tradition of building carrier based strike aircraft, notably the SBD Dauntless and the TBD Devastator torpedo bomber, both of which played vital roles in the early stages of the Pacific war.

The US Navy saw the Skyraider's successor as being a twin turboprop aircraft capable of performing close support and interdiction duties (with both conventional and nuclear weapons) and offering a maximum speed of 500mph (805km/h), a combat radius of 460 miles (740km), an offensive load of initially 1,000lb (454kg) then 2,000lb (907kg) and a maximum takeoff weight of about 30,000lb (13,608kg). In addition, and due to budget restrictions, a maximum unit price of $US1m was specified.

Douglas already had some experience with turboprops, via its XA2D-1 Skyshark which had first flown in 1950 and ordered into production. In many ways the obvious successor to the Skyraider (it bore a family resemblance and was intended for a similar role), the Skyshark was let down by its 5,525shp Allison XT-40 turboprop which comprised two T-38 engines coupled side by side and driving two three bladed contra-rotating propellers through a common gearbox.

Although offering good performance including a top speed of 525mph (849km/h) and climb rate of 7,200ft (2,194m)/min, the Skyshark's powerplant/propeller combination proved to be rather less than successful and development was abandoned after two prototypes had been built.

By early 1952 Ed Heinemann, the chief engineer at Douglas' El Segundo plant, had completed the overall concept of what would finally emerge as the A4D (or A-4 when the US rationalised its designation system in 1962) Skyhawk. Original proposals were based on a maximum takeoff weight of a mere 12,000lb (5,443kg) although this soon increased to 14,600lb (6,622kg) when the offensive load required was doubled and fuel of increased specific gravity (up from 6.0 to 6.5lb/US gallon) was introduced to the equation. The aircraft's empty weight was originally put at 8,136lb (3,690kg) but this had increased to 8,400lb (3,810kg) by the time the first aircraft flew, still a remarkably low figure considering the aircraft's capabilities.

Later Skyhawks typically weighed in at around 10,500lb (4,763kg) empty and the maximum weight increased to 24,500lb (11,113kg) for operations from carriers and 27,420lb (12,438kg) from land, still a remarkably 'compact' figure, especially considering that the final models were capable of lifting a 9,155lb (4,153kg) external load on the four underwing and single underfuselage pylons.

Heinemman specified a single, non afterburning Wright J65-W-2 single shaft turbojet of 7,200lb thrust to power the Skyhawk, although this variant was used only in the prototype. Pre production and early production Skyhawks had 7,700lb thrust J65-W-4 engines to cope with the aircraft's extra weight. The J65 was a licence built variant of the British Armstrong Siddeley Sapphire axial flow turbojet and was also used to power the Republic F-84F Thunderstreak, Martin B-57 Canberra, Grumman F11F Tiger, North American FJ Fury and other US types.

Efficiency of Design

From the start, Heinemann's intention was to keep his new design as simple and compact (and therefore light) as possible. Every component was analysed with this in mind, as was every design decision.

The short span cropped delta wing (a largish area was necessary to keep landing speeds down) measured only 27ft 6in (8.38m) from tip to tip, removing the need for wing folding and saving about 200lb (90kg) in the process; a single hydraulic system only was fitted to the prototypes and first production model; the internal fuel system was extremely simple with just one tank spread over the whole wing and another in the fuselage totalling 800US gallons (3,028 litres) – the lack of associated plumbing helped keep weight down; avionics were reduced from several of the in those days heavy 'black boxes' to single units with one connector where possible, and so on.

Structurally, the Skyhawk's wing was of all metal three spar design swept back at 33 degrees at quarter chord and bolted to the underside of the fuselage. The spars were machined from solid plates in one piece tip to tip and covered with one piece skins, and formed the single integral wing tank. In order to provide acceptable handling

An A-4B displays the characteristic 'inside out' rudder structure which was a feature of this and all subsequent Skyhawks. The A-4B entered service in 1957.

A pair of Royal Australian Navy A-4G Skyhawks. This variant was based on the A-4F but lacked that model's dorsal avionics 'hump'. It was also capable of carrying Sidewinder air-to-air missiles. (via Mike Kerr)

Skyhawk in its natural habitat – a TA-4J practicing an overshoot to a US Navy aircraft carrier. Note the lack of extended tailhook.

An A-4C currently flying in the USA as a 'warbird'. This Skyhawk model was the last to be powered by the Wright J65 (licence built) Armstrong Siddeley Sapphire) turbojet. (Jim Thorn)

The A-4E introduced a new powerplant in the 8,500lb thrust Pratt & Whitney J52-P-6 and the following A-4F had a 9,300lb thrust J52-P-8A.

characteristics, leading edge slats were fitted, while spoilers and split flaps were installed on the rear of the wing.

The Skyhawk's variable incidence tailplane was fitted with conventional elevators instead of the all moving tailplane which might have been expected; the single wheel main undercarriage units retracted forward into small fairings under the wing after first pre-shortening and turning through 90 degrees, the wheels themselves fitting into a cavity near the wing leading edge. The nosewheel also retracted forward and the undercarriage was hydraulically operated, as were the flaps, spoilers, air brakes and control surfaces.

The fuselage was of conventional all metal semi-monocoque construction with a detachable nose for access to the avionics mounted there, an airbrake on each side of the rear fuselage and a 'break' point in the centre so that when an engine change was needed, the whole rear fuselage and tail section detached for easy access. Skyhawks were built only as single seaters until 1965 when the first two seat operational trainer variant was produced.

Early Skyhawks had three stores hardpoints – one under each wing and another below the fuselage, and was each capable of carrying a variety of fuel tanks and bombs up to a typical combined capacity of 5,000lb (2,268kg), of which 3,000lb (1,361kg) could be carried under the fuselage. Internal armament was a pair of 20mm Colt Mk.12 cannon with 100 rounds each and mounted in the wing roots very close to the lower front fuselage.

Into Production

Douglas' proposals led to a contract for a prototype and 19 pre production aircraft in 1952, the first of which, designated the XA4D-1, flew on 22 June 1954 in the hands of pilot Bob Rahn. The first pre production A4D-1 took to the air for the first time on 14 August 1954.

Flight testing revealed the need for a revised tail cone design, vortex generators on the outer wings and a redesigned rudder structure to eliminate 'buzz'. These modifications would become standard on the later A4D-2 (A-4B) model, but in the meantime testing uncovered no serious problems.

Single and two seat Skyhawks of USN Squadron VA-125 parade for the camera.

Testing continued for two years before the Skyhawk entered service with the US Navy's VA-72 Attack Squadron at Quonset Point, Rhode Island, in September 1956. The aircraft was declared operational the following month and deliveries to the US Marine Corps (to VMA-224) began in January 1957.

In the meantime, the pre production aircraft had completed carrier trials (the first of which were on the USS *Ticonderoga* in September 1955) and set a world 500km closed circuit speed record of 695.127mph (1,118.7km/h) at Edwards Air Force Base.

Production of the initial A-4A model (for convenience we'll use the current designations from here) amounted to 165 examples including the 19 pre production aircraft which were at first designated YA4D-1s. Initially powered by the 7,700lb thrust J65-W-4 engine, they were updated over the years and ended up with 8,500lb thrust J65-W-18 engines installed, similar to those fitted to the later A-4E.

The efficiency of this first production model Skyhawk can be assessed to some extent by looking at its empty to maximum weight ratio: 8,400lb (3,810kg) to 20,000lb (9,072kg) or 2.38 to 1, an extraordinary figure and indicative of Heinemann's success in achieving his goal of 'minimisation'.

The subsequent A-4B and C models also had 7,700lb thrust J65s, and like their predecessor, were progressively upgraded to the more powerful engine.

Developing the Breed

Five hundred and forty-two A-4Bs (A4D-2) followed on the production line, the first of them flying in March 1956 and entering service the following year. Improvements over the first model included the introduction of a stiffer single surface rudder (with its characteristic 'inside out' central skin and external stiffeners design), powered flying controls, dual hydraulic systems, single point pressure refuelling, a strengthened structure, an improved bomb delivery system, provision to carry the Martin Bullpup air-to-surface missile (ASM), an automatic dead reckoning navigation computer and provision for inflight refuelling.

The latter included development of a 'buddy' refuelling pack for the underfuselage station incorporating a 300US gallon (1,135 litre) tank and the necessary drogue and hose equipment. All of the A-4B's external fuel and half of its internal fuel could be transferred to other aircraft when it was used as a tanker.

As noted above, the A-4B's initial powerplant was the 7,700lb thrust J65-W-16, but aircraft were progressively modified to accept 8,500lb thrust J65-W-18 engines.

In January 1959 a pair of A-4Bs flown by Major J P Flynn and Captain C D Warfield of the 2nd Marine Aircraft Wing flew non stop from El Toro, California, to Cherry Point, North Carolina, – a distance of 2,082 miles (3,350km) – in 4hr 25min without the use of inflight refuelling.

The next model, the A-4C (A4D-2N), was similarly powered and subsequently modified and was the most numerous of all the Skyhawk variants with 638 manufactured. The first A-4C flew in August 1958, deliveries began in December 1959 and the final example was handed over in December 1962. At the peak of its career, the A-4C equipped 23 US Navy and nine US Marines squadrons plus training and other miscellaneous units.

The A-4C improved the Skyhawk's operational effectiveness which had hitherto been limited to daylight attack missions by providing some poor weather and light capability with installation of the very compact and lightweight Westinghouse APG-53A radar. The radar weighed only 90lb (41kg), used a dish of just 16in (40.6cm) diameter and apart from the usual surface mapping and ranging functions, was capable of providing terrain clearance information for low altitude, high speed sorties.

Other mission enhancement equipment installed in the A-4C included blind and low altitude bombing systems, an all attitude reference and loft bombing system, an improved gunsight and an automatic flight control system. The A-4C weighed in at 9,728lb (4,413kg) empty and differed externally from its predecessors by having a slightly lengthened nose to accommodate the new radar and other equipment.

New Engine, More Capability

The A-4E Skyhawk represented the biggest single redesign of the family, the fundamental change being the installation of the Pratt & Whitney J52-P-6 two spool turbojet of 8,500lb thrust in place of the original J65

Skyhawks have long been used by the US Navy for its adversary air combat training programme. By 1993 earlier models were being replaced by F/A-18 Hornets in this role, giving the 'Top Guns' a rather faster opponent to do battle with.

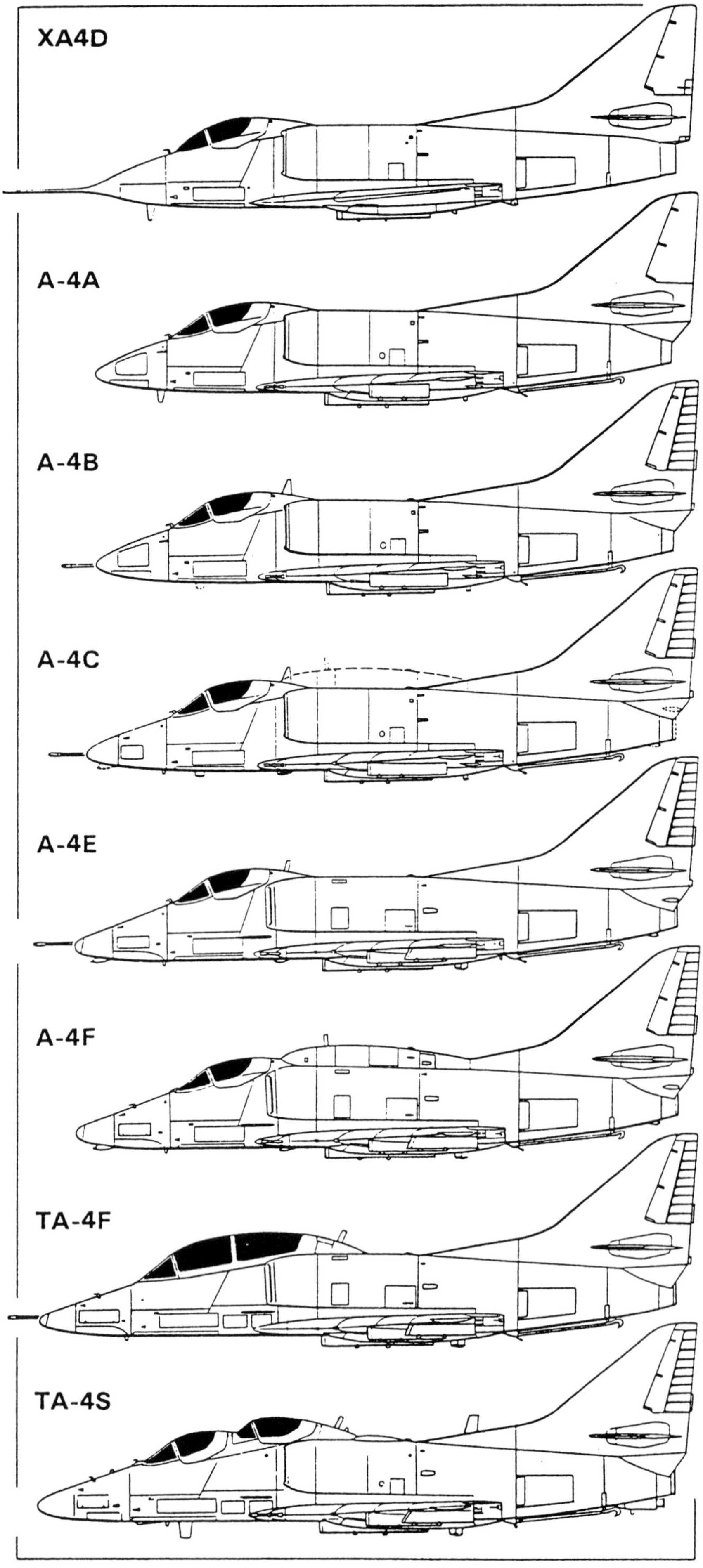

Sapphire. The new engine (which also powers the Grumman A-6 Intruder and EA-6 Prowler) offered greater growth potential and better fuel economy than the J65, which was nearing the end of its development. The first of two evaluation YA-4Es (YA4D-5s) flew in July 1961 and was followed by 494 production A-4Es, the first of them being handed over in November 1962. Production ended in April 1966.

The empty weight of the A-4E grew to 9,853lb (4,469kg) and the maximum weight to 24,500lb (11,113kg), but with that came a substantial 27% increase in payload/range capability and the capacity to carry a theoretical maximum of 9,155lb (4,153kg) in external stores. This was due to the addition of an extra pair of underwing hardpoints making four in all plus the underfuselage centreline point.

A normal maximum load of 8,200lb (3,719kg) was carried, comprising up to 20 separate items, while supporting these changes were a further strengthened structure (29 per cent of it was redesigned), a Douglas 'Escapac' zero height/90 knots rocket ejection seat, Doppler radar, Tacan, radar altimeter and improved low altitude bombing system.

Of interest, and indicative of the normal development associated with such an aircraft, is the fact that the avionics installation weighed 526lb (239kg) in the A-4E compared with just 138lb (63kg) in the A-4A.

By mid 1965 the Skyhawk had seen enough service to properly evaluate its record. By then the type had amassed more than 1,700,000 flying hours and had established the best reliability record of any US Navy attack aircraft with a mean time between failures serious enough to cancel or fail to complete a mission of 23.6 hours. Using the standard formula of the day, this meant the Skyhawk had a 90 per cent probability of completing a 2.5 hour mission.

Maintenance records showed an average 12.8 maintenance man hours per flight hour – outstanding for a carrier based aircraft at the time – while on active service flying hazardous close support and interdiction missions in Vietnam, the Skyhawk displayed an availability rate of about 95 per cent, an enviable record.

The A-4F was the final major single seat Skyhawk variant developed for the US Navy. First flight was recorded in August 1966 and delivery of 146 examples began in June 1967 (two months after the McDonnell Aircraft Corporation and the Douglas Aircraft Company merged to form McDonnell Douglas) for completion the following year. The A-4F continued the Skyhawk's weight growth, coming in at 10,448lb (4,739kg) empty, of which the avionics contributed 744lb (337kg). Maximum weights remained as before: 24,500lb (11,113kg) for carrier operations and 27,420lb (12,438kg) from land.

It was extra avionics which contributed to the A-4F's most obvious external change over its predecessors; the dorsal hump behind the cockpit, designed to carry them and a legacy of the aircraft's compact size and therefore limited internal space.

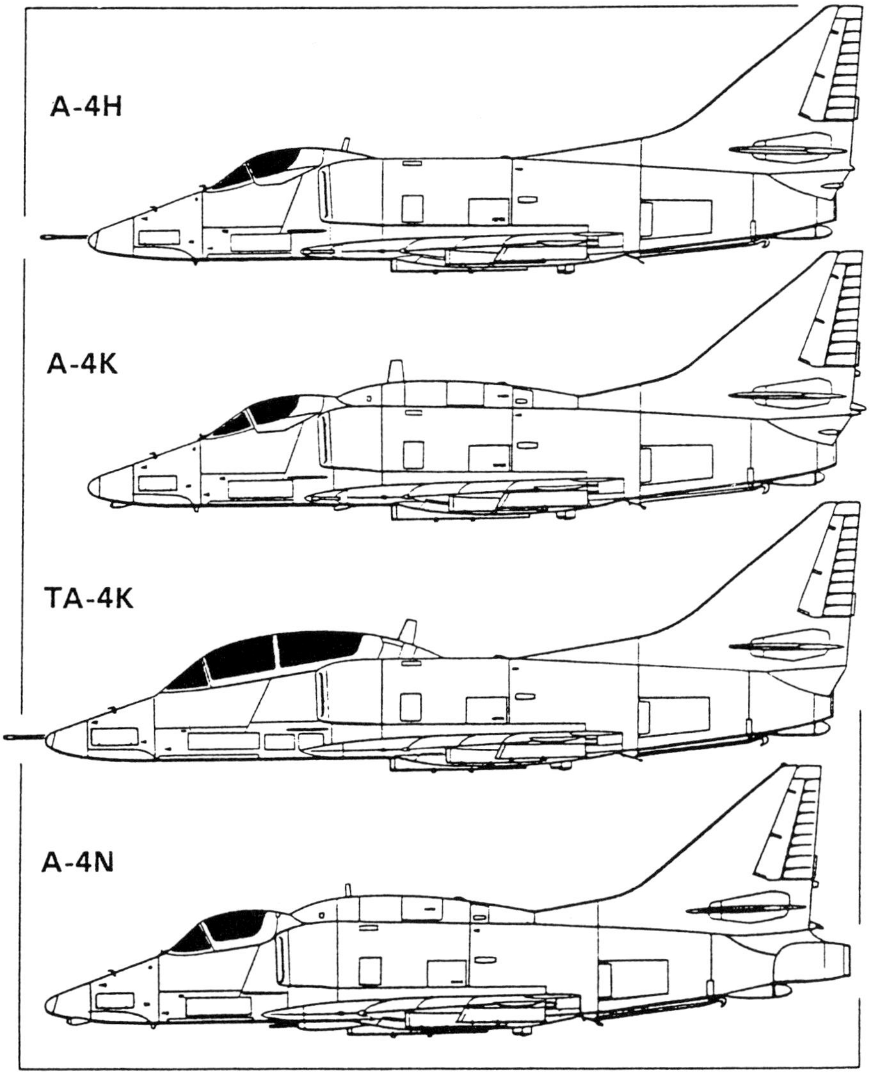

McDONNELL DOUGLAS A-4G SKYHAWK

Powerplant: *One Pratt & Whitney J52-P-8A two-spool turbojet rated at 9,300lb static thrust; fuel in integral wing tanks and fuselage tank aft of cockpit, combined capacity 800US gal (3,028 l); maximum external fuel capacity 1,200US gal (4,545 l) comprising range of tanks on centreline and two underwing stations; flight refuelling probe on starboard side of nose; provision for Douglas self contained flight refuelling unit on centreline station.*

Dimensions: *Wing span 27ft 6in (8.38m); length (excluding flight refuelling probe) 40ft 3.25in (12.27m); height 15ft 2in (4.62m); wing area 260sq ft (24.15sq m); wheel track 7ft 9.5in (2.37m).*

Weights: *Max t-o (carrier ops) 24,500lb (11,113kg), max t-o (land ops) 27,420lb (12,438kg); typical empty 10,100lb (4,581kg).*

Avionics: *(1967) ASN-1 Dead Reckoning Navigation Computer; APG-53A terrain clearance, ground mapping, air-to-ground ranging, shipping search and weather mapping radar (limited air-to-air capability); Bendix Automatic Flight Control System; UHF Direction Finder; TACAN; radar altimeter; UHF radio with 20 preset channels and up to 1,750 manually selectable frequencies; IFF (Identification Friend or Foe) equipment; intercom on TA-4G.*

Armament: *Two 20mm Colt Mk.12 cannon in wing roots with 100rpg; provision for two underwing AIM-9B Sidewinder heat seeking air-to-air missiles. Five stores points: one under fuselage and two under each wing with typical combined capacity of 8,000lb (3,629kg). Typical RAN loads: clusters of three 250 or 500lb (113 or 227kg) at three points; single 250 or 500lb (113 or 227kg) bombs at five points; single 1,000lb (454kg) bomb at three points; 5in Zuni or 2.75in FFAR rocket pods at five points; parachute flares.*

Performance: *Max speed (clean) 675mph (1,086km/h) or Mach 0.88 at sea level, 612mph (985km/h) or Mach 0.8 at 35,000ft; Max speed (external stores) 610mph (982km/h) or Mach 0.8 at sea level, 610mph (981km/h) or Mach 0.85 at 30,000ft; typical cruise 490mph (788km/h); initial climb rate 5,620ft (1,713m)/min; service ceiling (clean) 47,900ft (14,600m); combat radius (hi-lo-hi mission with 4,000lb stores) 400 miles (644km); ferry range with max external fuel 2,440 miles (3,926km).*

Early A-4Fs lacked this feature but were retrofitted, as were many A-4Es.

Under the A-4F's skin were several significant changes: a more powerful 9,300lb thrust J52-P-8A engine, cockpit armour and generally improved protection from ground fire, a steerable nosewheel, lift spoilers on the wings to shorten the landing run, Douglas Escapac 1C3 zero-zero ejection seat and low pressure tyres.

The eight A-4G Skyhawks delivered to the Royal Australian Navy in 1967 were basically A-4Fs without the dorsal avionics hump and with provision for the carriage of Sidewinder air-to-air missiles. The further eight A-4Gs delivered in 1971 were ex-USN A-4Fs refurbished and modified to Australian standards. The RAN's Skyhawks will be discussed in more detail in the following chapter.

Proposals and Two Seaters

Several export variants of the Skyhawk were proposed in the mid 1960s, one of them incorporating another change of engine, this time to the Rolls-Royce Spey turbofan, the particular non afterburning variant considered offering 12,000lb thrust.

The Spey-Skyhawk proposal followed an unsuccessful bid in 1965 to sell the aircraft to Canada under the designation CA-4E (single seater) which featured the engine and other improvements incorporated in the A-4F which was to follow, and the CA-4F two seater. Douglas then looked at the re-engined Skyhawk with a view to offering it to various European nations including Belgium, The Netherlands and Italy.

None ordered it and not even a prototype was built, but had it entered production the Spey-Skyhawk was to have featured a slightly deeper and longer fuselage with internal structural modifications to accommodate the larger and heavier turbofan, larger air intakes, a fin of increased area and rearrangement of equipment in the centre and rear fuselage. Despite its low purchase price ($US764,000 with all-weather navigation equipment), no customers were found for the aircraft.

A Skyhawk variant which did go ahead was the tandem two seater dual control trainer, originally developed to replace the US Navy's Grumman TF-9J Cougar. An order for two TA-4E prototypes was placed in 1964 and the first example flew in June 1965.

Compared with the single seater, the Skyhawk trainer had a lengthened (by 2ft 4in/71cm) front fuselage incorporating two seats in tandem under a single canopy, the space taken by the rear seat reducing the aircraft's internal fuel capacity by 136 US gallons to 664 US gallons (2,513 litres). The rear seat was raised 9in (23cm) compared with the front.

Because the trainer took on most of the features of the A-4F (P-8A engine, steerable nosewheel, lift spoilers, zero-zero seats etc), it was redesignated TA-4F shortly after its first flight. It retained operational capability with the A-4F's five weapons hardpoints but had reduced avionics capability.

(top) Two dual seat TA-4 Skyhawk versions were produced for the US Navy, the TA-4F and less powerful and simplified J, as illustrated here. These aircraft are from TW-2 on board the USS Lexington. *(centre) New Zealand's 14 A-4Ks and TA-4Ks were delivered in 1970 and featured the engine of the A-4F with some of the features of the later A-4M including a braking parachute and larger fin and rudder. Ten former RAN A-4Gs and TA-4Gs were acquired in 1984 and upgraded to 'K' specifications before being further modified under the* Kahu *programme. (bottom) A Singaporean A-4S, a greatly upgraded ex US Navy A-4B. Further development has resulted in the A-4S-1 Super Skyhawk powered by a non afterburning GE F404 turbofan. (Paul Merritt/Bill Lines)*

Singapore's TA-4S two seaters were converted from single seat A-4Bs and feature separate cockpit canopies rather than the usual large single piece unit. (Bill Lines)

Deliveries of the TA-4F began in May 1966 and a total of 242 was built plus two similar TA-4Gs for the Royal Australian Navy delivered in 1967 (a further pair were delivered in 1971, converted from US Navy TA-4Fs). As noted below, other customers for new build TA-4 Skyhawks were Israel, New Zealand and Kuwait.

The second major Skyhawk trainer variant was the TA-4J, 291 of which were built from 1969. A simplified version of the TA-4F for the US Navy's advanced flying training units, the 'J' has 8,500lb thrust J52-P-6 engines and no nav-attack system, although when built the aircraft retained provision for this equipment.

Apart from Australia, there have been several other export customers for the A-4B-F model Skyhawks. The first was Argentina, which received 25 former US Navy A-4Bs for its air force in 1966, 25 more in 1970 and another 25 A-4Cs in 1971. The A-4Bs were redesignated A-4Ps, while a batch of 16 refurbished A-4Bs (designated A-4Q) went to the Argentine Navy in 1971.

New production Skyhawks based on the A-4F and TA-4F were delivered to Israel (90 A-4Hs and 10 TA-4Hs from 1967) and New Zealand (10 A-4Ks and four TA-4Ks in 1970).

The Israeli aircraft differed from the standard specification in several ways including the installation of two 30mm DEFA cannon in the wing roots (with 150 rounds per gun) in place of the standard 20mm units, a braking parachute and the larger fin which would later appear on the A-4M and -N 'Skyhawk II' models referred to below.

The Royal New Zealand Air Force's A-4Ks and TA-4Ks also featured a braking parachute and modified fin and were later supplemented by the eight surviving Royal Australian Navy A-4Gs and two TA-4Gs in 1984. These were modified to 'K' standards and then to the substantially upgraded *Kahu* specification after that. New Zealand's Skyhawks are discussed in greater detail in a later chapter.

Other customers for second-hand Skyhawks have included Indonesia, Malaysia and Singapore, the former acquiring aircraft from Israel and the USA and the latter 46 ex-USN A-4Bs converted to A-4S and two seat TA-4S standards, initially by Lockheed Air Services and then in Singapore.

The first conversions were performed in 1973 and included the installation of 8,100lb thrust J65-W-20 engines, 30mm Aden cannon, modernised nav-attack systems, a braking parachute, a Ferranti lead-computing gunsight and lift spoilers. The seven TA-4S conversions are unusual in that separate canopies cover the two seats rather than the usual one piece arrangement.

Singapore subsequently acquired other Skyhawks for conversion to A-4S standards and in 1986 completed the first two

The A-4N for Israel was a version of the M with an improved nav-attack system and modified cockpit layout. First flown in July 1972, 117 were delivered. (MDC)

conversions of its own aircraft to A-4S-1 'Super Skyhawk' specifications, the most advanced Skyhawk modification programme as yet undertaken.

The conversion involves replacement of the standard engine with a non afterburning General Electric F404-GE-100D turbofan of 10,800lb thrust. In combination with avionics and systems updates, the A-4S-1 offers a 15 per cent increase in dash speed and substantial improvements in acceleration, turn rate and takeoff performance. Fifty-two conversions were carried out initially, with more to come, and the work has been carried out with the assistance of Grumman and General Electric.

Other designations given to earlier model Skyhawks include A-4L, A-4Cs converted to A-4F avionics standards but with original engines and three stores positions retained.

Kuwait received 30 A-KU and six TA-4KUs in 1977. In the Gulf War Kuwaiti Skyhawks operated from a road near their bombed airfield after the Iraqi invasion.

New Generation

In early 1964, the Vought Corporation won a competition for a new naval attack aircraft for the US Navy and Marines which would ultimately replace the Skyhawk. The resulting A-7 Corsair II was a larger and more capable aircraft than its smaller predecessor, but considerably more expensive and maintenance intensive.

The Navy ordered Corsairs but the Marines opted to stay with the Skyhawk, funding the final generation, the A-4M or 'Skyhawk II' as it was dubbed by its manufacturer.

Externally, the A-4M differed from earlier models in having a larger canopy for a better view, the larger, squarer fin previously seen on the A-4H and A-4K and an angled flight refuelling probe.

A more powerful 11,200lb thrust J52-P-408 engine gave a vastly improved rate of climb and reduced takeoff length while other modifications included installation of a braking parachute and twice the gun ammunition (200 rounds each) of the earlier models. The electrical generator was uprated and the aircraft featured a self contained gas turbine engine starter and wind driven ram air turbine for back up power generation. Empty weight was 10,800lb (4,899kg) and maximum weights remained as before. An overall combat efficiency improvement in the order of 30 per cent was claimed.

The first A-4M flew in April 1970 and deliveries to the USMC began a year later. Production ran to 162 examples, the last of which (delivered in February 1979) was also the 2,960th and final Skyhawk to leave the production line.

Variants of the A-4M were exported to Israel and Kuwait. The Israeli version was the A-4N, first flown in July 1972 and featuring an improved nav-attack system and modified cockpit layout. Some Israeli A-4Ns have undergone further modification including the fitting of an extended tailpipe to reduce their infra-red signature and thus their vulnerability to heat seeking missiles.

Israel received 117 A-4Ns, bringing its procurement of Skyhawks to no fewer than 294, or 10 per cent of the total production run.

The other purchaser of Skyhawk IIs was Kuwait, which received 30 A-4KUs and six TA-4KUs in 1977.

A-4M 'Skyhawk II', the final major production version developed for the US Marine Corps with more power, a braking parachute, increased gun ammunition, a larger fin and rudder, larger canopy and about 30 per cent greater operational efficiency than its immediate predecessor. The 2,960th and last Skyhawk was an A-4M delivered in February 1979.

SKYHAWKS FOR THE RAN

Despite a long tradition of naval aviation dating back to December 1917 when Flt Lt F N Fox performed the successful takeoff of a Sopwith Pup from the quarterdeck of the cruiser HMAS *Australia*, it wasn't until 28 August 1948 when the Royal Australian Navy Fleet Air Arm as a separate identity was established.

On that day, the 20th Carrier Air Group was commissioned at Eglington Naval Air Station, England, equipped with Hawker Sea Furies and Fairey Fireflies. In 1950, a second Carrier Air Group (the 21st) was established, also equipped with Sea Furies and Fireflies.

The Fleet Air Arm was established to provide a force capable of providing aerial anti submarine and fighter protection for the fleet and was based around two aircraft carriers and shore establishments, the latter comprising former RAAF airfields at Nowra on the south coast of New South Wales and Schofields west of Sydney. These were named HMAS *Albatross* (after the seaplane carrier which served the RAN during the 1930s) and HMAS *Nirimba*, respectively.

The RAN's Carriers

The first of three light fleet aircraft carriers operated by the Royal Australian Navy was HMAS *Sydney*, the former Royal Navy *Majestic* class ship HMS *Terrible* of 14,744 tonnes standard displacement. She was transferred to the RAN in 1948 and saw active service in the Korean war, her Sea Furies and Fireflies logging 2,366 sorties against the enemy between September 1951 and January 1952.

Sydney retained a straight through deck (rather than an angled one) in her career with the RAN, making her unsuitable for the operation of jet aircraft. As a result, and due to government policy which decided a single carrier was sufficient for Australia's operational needs, *Sydney* was relegated to training duties in 1955 and served in that role until 1958 when she was put into 'special reserve' – mothballed, but capable of being returned to service.

This happened four years later, following a 1960 decision and when tensions associated with the period known as 'confrontation' with Indonesia began, these in turn resulting from the formation of the Federation of Malaysia, a coalition of the former British colonies Malaya, Singapore (which later withdrew), Sarawak (North Borneo) and Sabah.

Sydney was therefore recommissioned in 1962 as a fast troop transport, combining this role with training and making numerous voyages to Vietnam in support of Australia's substantial involvement in that particular war.

Sydney was finally decommissioned in 1972 and sold for scrap in 1975.

The second of the RAN's carriers, the *Colossus* class HMAS *Vengeance*, served only briefly on loan from the Royal Navy pending the arrival of HMAS *Melbourne*.

Vengeance was commissioned into RAN service in November 1952 and was returned to Britain in August 1955. While in RAN service, the carrier operated Sea Furies, Fireflies and Bristol Sycamore helicopters.

HMAS *Melbourne*

Perhaps the best known of the RAN's aircraft carriers was HMAS *Melbourne*, the Flagship of the Fleet from her commissioning in October 1955 to her paying off in June 1982. In the meantime she operated Sea Venom, Gannet, Skyhawk and Tracker fixed wing aircraft along with Wessex and Sea King anti submarine helicopters.

Laid down in 1943 as the nameship of the Royal Navy's *Majestic* light fleet carrier class, *Melbourne* (as she was subsequently named) was launched at Barrow-in-Furness, England, in February 1945. With the end of World War II, work on the ship stopped but resumed when the Australian government in 1947 decided to purchase her

Royal Australian Navy A-4G Skyhawk N13-154905/884 from the first batch of 10 aircraft delivered in 1967.

HMAS Melbourne *photographed in 1980 with Skyhawks, Trackers and Sea Kings on the flight deck.* Melbourne *started life as HMS* Majestic *but was completed for the Royal Australian Navy after extensive modification including the fitting of an angled flight deck, steam catapult, new arrestor gear and mirror deck landing system. (via Ross Gillett)*

(along with *Sydney*) for the Royal Australian Navy. The combined cost of the two carriers was £11,976,090 ($23,952,180) plus a maintenance cost of £11,432,400 ($22,864,800).

Before commissioning into RAN service, the 16,256 tonnes standard displacement *Melbourne* underwent extensive redesign and modification to enable her to operate the latest larger and heavier jets and other types of aircraft. This work included the lengthening of the flight deck and incorporating a 5.5deg angle, along with the fitting of larger flight deck lifts, a steam catapult, new arrestor gear and mirror deck landing system.

Melbourne was commissioned on 28 October 1955 and after working up in British waters with her new Sea Venom and Gannet aircraft on board she sailed to Australia under the command of Capt G G O Gatacre, arriving at Sydney in April 1956.

The increasing importance of anti submarine warfare (ASW) was indicated in 1963 by the inclusion of the Westland Wessex helicopter in the ship's aircraft complement, although this was symptomatic of greater philosophical changes as to what the RAN Fleet Air Arm's role should be, changes which very nearly saw the end of fixed wing flying from *Melbourne* some two decades before it finally happened.

Had those changes been implemented when planned, the Royal Australian Navy would not have purchased the A-4 Skyhawk (or the S-2 Tracker) and fixed wing operations would have ceased in 1963.

HMAS MELBOURNE
Milestones: *Laid down 15 April 1943 (as HMS Majestic); launched 28 February 1945; ordered by RAN June 1947; commissioned (as HMAS Melbourne) 28 October 1955; paid off 30 June 1982.*
Builder: *Vickers-Armstrong (Barrow-in-Furness, Cumbria, England).*
Displacement: *16,256 tonnes standard; 20,645 tonnes deep load.*
Dimensions: *overall length 701ft 6in (213.82m); waterline length 650ft (198.12m); flight deck length 690ft (210.31m); beam (hull) 80ft (24.38m); beam (flight deck) 105ft (32.0m); draught (deep load) 25ft (7.72m).*
Machinery: *Two Parsons single reduction geared turbines, four Admiralty 3-drum boilers; 42,000 shaft horsepower.*
Performance: *Max speed 24.5 knots (45.4km/h); endurance 12,000nm (22,200km) at 14 knots (26km/h).*
Bunkerage: *Fuel oil 2,948 tonnes; diesel 356 tonnes; Avtur 140,171gal; Avgas 3,744gal.*
Armament: *(early) six twin and 13 single 40mm Bofors guns; (late) four twin and four single 40mm Bofors guns.*
Accommodation: *Crew of 1,210; aircraft 22-27. Sea Venom and Gannet (1956), Wessex (1963), A-4 Skyhawk and S-2 Tracker (1967), Sea King (1975).*

Fleet Air Arm Threatened

In September 1959 the Australian government announced that the Royal Australian Navy as a whole would concentrate on anti submarine warfare operations and that the defence forces generally would be standardised with those of the United States.

Among the various proposals as they directly affected the Fleet Air Arm was one that it be disbanded in 1963 when the Gannets and Sea Venoms reach the end of their planned operational lives. It was considered that any likely replacement aircraft would not be able to operate from *Melbourne*.

This was modified slightly in 1960 when it was decided that *Melbourne* would continue in service but converted for anti submarine duties equipped with Wessex helicopters. For the moment, the decision to end fixed wing flying in 1963 remained in place.

This situation changed that very year when the threats associated with 'confrontation' with Indonesia were fully appreciated and the Sea Venoms and Gannets were given the go ahead to continue operations from *Melbourne* until 1967.

Ring in the new ... an RAN Sea Venom poses with its successors, an A-4G and TA-4G Skyhawk. (via Ross Gillett)

(left) A Skyhawk being hoisted aboard HMAS Melbourne *at North Island Naval Air Station, San Diego California in October 1967 preparatory to delivery to Australia. The voyage brought not only the first 10 RAN Skyhawks to Australia but also 14 new Grumman S-2 Trackers. (right) RAN Skyhawks of the first batch on board* Melbourne *before leaving the USA for Australia. (via Mike Kerr/Ross Gillett)*

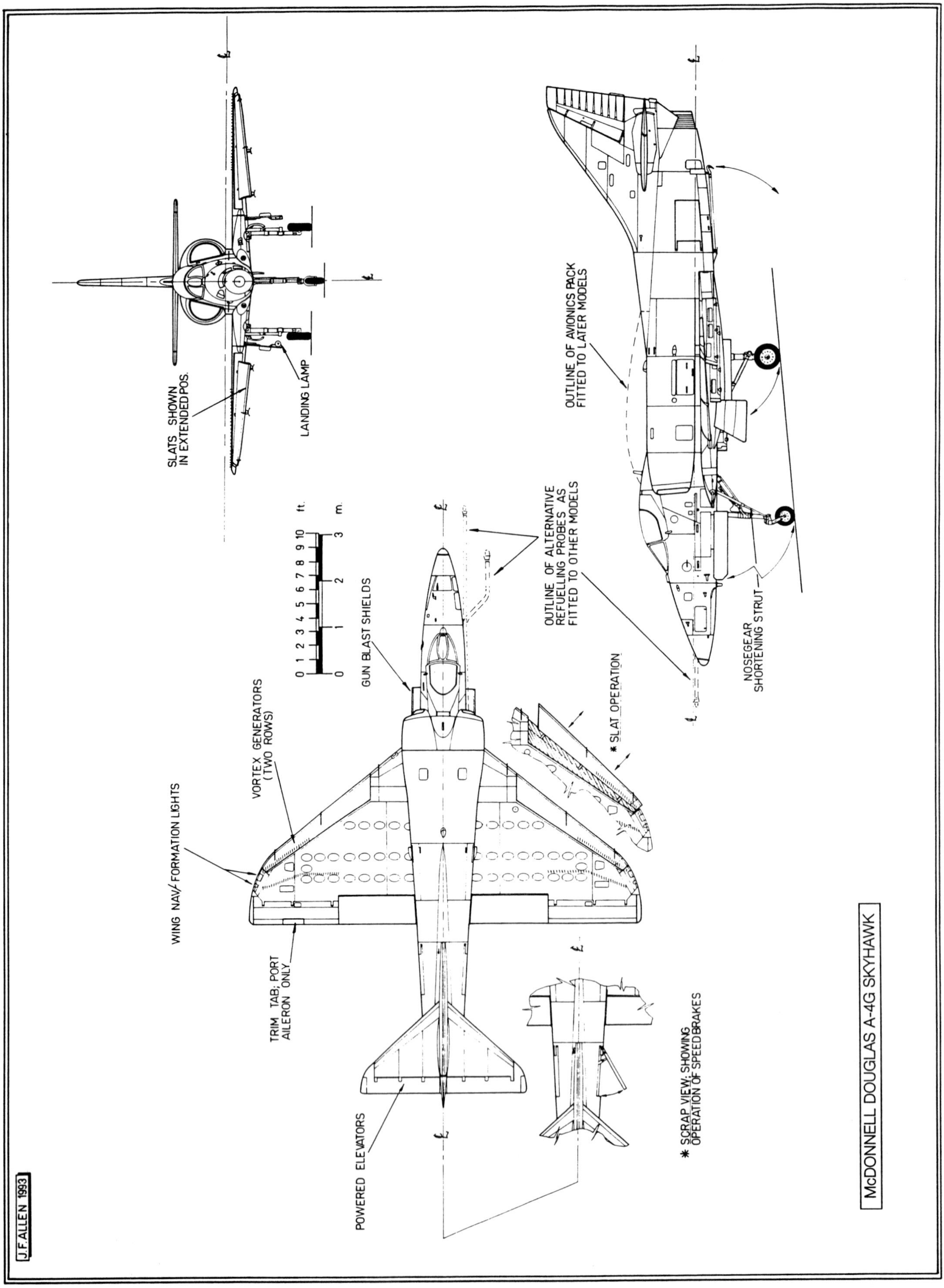
McDONNELL DOUGLAS A-4G SKYHAWK
J.F.ALLEN 1993
SLATS SHOWN IN EXTENDED POS.
LANDING LAMP
OUTLINE OF AVIONICS PACK FITTED TO LATER MODELS
OUTLINE OF ALTERNATIVE REFUELLING PROBES AS FITTED TO OTHER MODELS
NOSEGEAR SHORTENING STRUT
ft.
0 1 2 3 4 5 6 7 8 9 10
m.
0 1 2 3
GUN BLAST SHIELDS
VORTEX GENERATORS (TWO ROWS)
WING NAV/FORMATION LIGHTS
* SLAT OPERATION
TRIM TAB; PORT AILERON ONLY
POWERED ELEVATORS
* SCRAP VIEW; SHOWING OPERATION OF SPEEDBRAKES

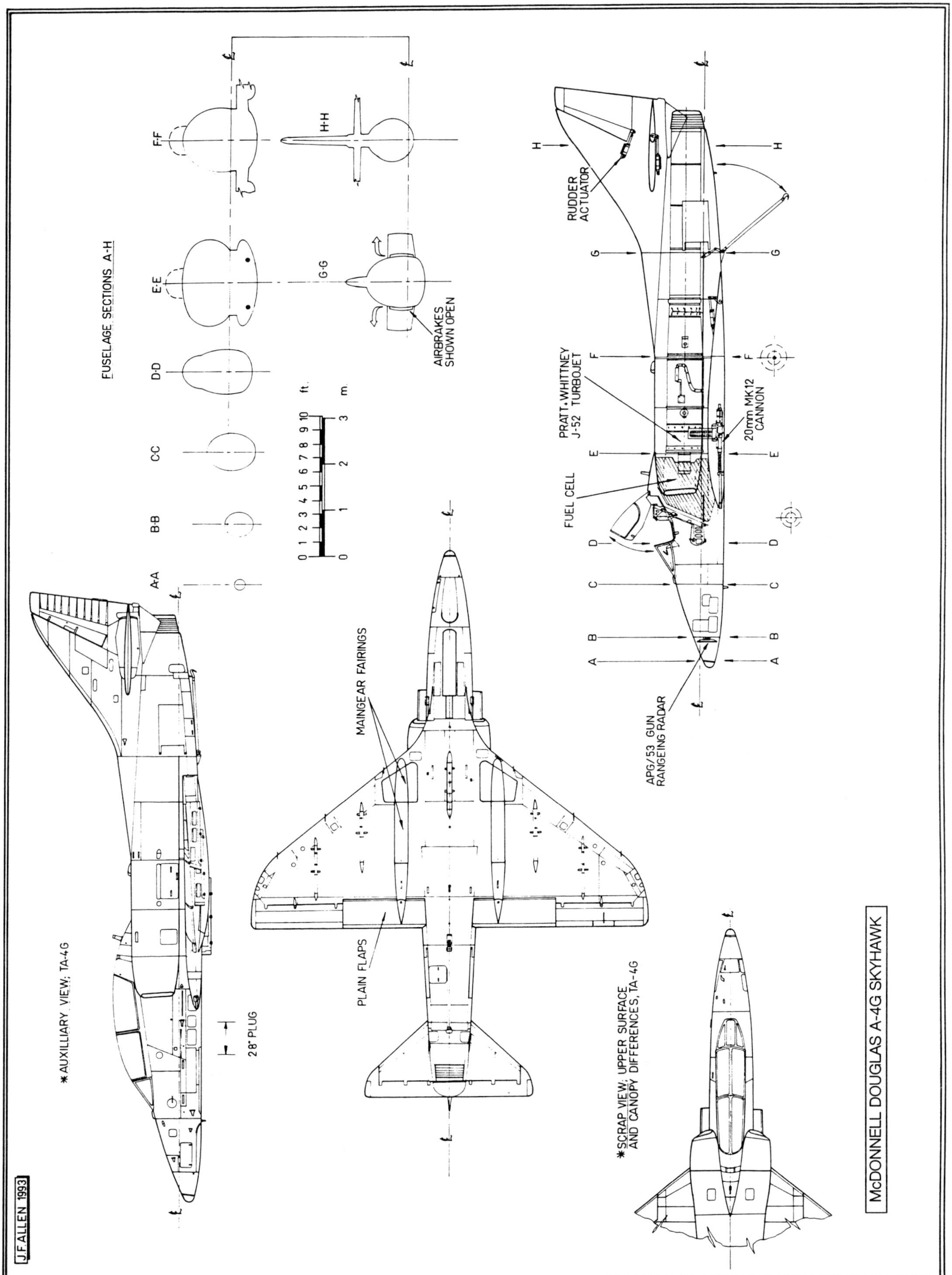
McDONNELL DOUGLAS A-4G SKYHAWK
J.F.ALLEN 1993
FUSELAGE SECTIONS A-H
A-A
B-B
C-C
D-D
E-E
F-F
G-G
H-H
AIRBRAKES SHOWN OPEN
ft.
m
RUDDER ACTUATOR
PRATT·WHITTNEY J-52 TURBOJET
20mm MK12 CANNON
FUEL CELL
APG/53 GUN RANGEING RADAR
MAINGEAR FAIRINGS
PLAIN FLAPS
✱AUXILLIARY VIEW; TA-4G
28" PLUG
✱SCRAP VIEW; UPPER SURFACE AND CANOPY DIFFERENCES, TA-4G

Hooks down, a trio of 805 Squadron Skyhawks overfly HMAS Melbourne. *A solitary Tracker occupies the forward flight deck.*

(left) The first RAN Skyhawks arrived at Jervis Bay NSW in November 1967 where they were transferred by barge to the shore and then by road to their base at nearby Nowra. This aircraft is N13-154905/884. (right) TA-4G Skyhawk N13-154911/880, also from the 1967 batch. 880 was one of the surviving RAN Skyhawks which went to the Royal New Zealand Air Force in 1984. (via Ross Gillett/DoD)

Skyhawks Ordered

This same situation resulted in a complete reassessment of the Fleet Air Arm's worth, with the result of important new orders for fixed wing aircraft in the mid 1960s: 14 Grumman S-2 Tracker twin piston engined anti submarine aircraft in November 1964 and 10 Douglas A-4 Skyhawk jet fighter-bombers in October 1965 to replace the Gannets and Sea Venoms.

The 1963-65 period saw the placing of substantial aircraft orders for the Australian Defence Forces generally. Apart from the Trackers and Skyhawks for the RAN, the RAAF also benefitted from an equipment acquisition programme which included F-111Cs, more Iroquois helicopters, Caribou tactical transports, an extra squadron of C-130 Hercules transports, P-3 Orion ASW aircraft, Macchi MB.326 jet trainers, HS.748 navigation trainers and BAC One-Eleven and Dassault Falcon VIP aircraft.

The 10 Skyhawks were listed as A-4E variants and were costed at £9.2m ($18.4m) for the programme. Deliveries were scheduled to begin in 1967 and the Skyhawks would serve on *Melbourne* alongside the ASW Trackers and Wessex helicopters.

The Skyhawks were later reidentified as A-4Gs, based on the US Navy's A-4F but incorporating some modifications to meet Australian requirements. Of the 10, eight were single seat A-4Gs and the remaining pair two seat TA-4G trainers.

The A-4G retained the strike capabilities of its American counterparts but was intended as much for fleet defence duties when embarked on *Melbourne*. As a consequence, they were capable of being fitted with a pair of AIM-9B Sidewinder infra-red air-to-air missiles (the same armed the RAAF's Mirages) but could also carry 250lb (113kg) or 500lb (227kg) bombs, 2.75 or 5in rocket pods and other stores for use in maritime strike, army support and fleet support missions. Changes were made to the avionics fit and externally the Australian Skyhawks differed from the American A-4F by the lack of the distinctive dorsal avionics 'hump'.

Interestingly, the TA-4Gs were not able to be operated from the *Melbourne*, the changed centre of gravity in the two seater resulting in slower elevator response at low speed, this in turn preventing the TA-4G from lifting its nosewheel from the deck of the relatively small carrier in time in the case of a 'bolter' (missing the arrestor wires and going around), or so it was calculated.

According to the RAN the Skyhawk had: "proved itself in combat in Vietnam as rugged, easily maintainable and second to none in cost effectiveness". All of which was true. The Skyhawk "will provide the carrier with a proven counter against hostile reconnaissance aircraft and at the same time provide a limited strike capacity against a surface force attack", the Navy went on.

Delivery and Second Order

The first two A-4Gs (one of each model) were handed over to the Royal Australian Navy at the Douglas (or McDonnell Douglas, as it now was) facility at Long Beach, California, on 26 July 1967.

Attending the ceremony were officials of the Australian government, RAN, USN, USMC and US Coast Guard officers and representatives from Long Beach city and the manufacturer. The log books for the two aircraft were handed to Vice Admiral Allen M Shinn USN (commander of the Naval Force, US Pacific Fleet) by Donald W Douglas Jr; they were then presented to Rear Admiral G J B Crabb CBE DSO, head of the Australian Joint services Staff in Washington DC.

All ten aircraft arrived in Australia (at Jervis Bay NSW from where they were transported by road to their base at nearby Nowra) during November 1967 aboard HMAS *Melbourne*. The carrier had sailed to the USA to collect the Skyhawks and the Trackers prior to a major refit which would enable her to effectively operate the new aircraft. The carrier's previous equipment – the Sea Venoms of 805 Squadron and Gannets of 816 Squadron – had meanwhile been withdrawn from front line service a few months earlier.

The Skyhawks were collected from North Island Naval Air Station at San Diego, California, in October 1967 where they were sealed and covered in protective material before being hoisted aboard the carrier.

RAN Skyhawk operations are discussed in the following chapter, but in the meantime a second order for ten aircraft (again eight single seaters and two trainers) was placed in March 1970. These aircraft were A-4Fs and TA-4Fs purchased from USN stocks and modified to A/TA-4G standards. They were delivered to Australia in August 1971, unloading at Jervis Bay after a voyage from the USA on board the former aircraft carrier and then troop transport, HMAS *Sydney*.

Two RAN Skyhawks from the second batch of 10 delivered in August 1971. The eight A-4Gs and two TA-4Gs were 'F' models purchased from USN stocks and modified to Australian specifications. Pictured are A-4G N13-155052/871 and TA-4G N13-154647/878.

RAN SKYHAWKS Serial and Side Numbers		
1st Batch (deliv 11/67)		
Serial No	**Side**	**Model**
N13-154903	882	A-4G
N13-154904	883	A-4G
N13-154905	884	A-4G
N13-154906	885	A-4G
N13-154907	886	A-4G
N13-154908	887	A-4G
N13-154909	888	A-4G
N13-154910	889	A-4G
N13-154911	880	TA-4G
N13-154912	881	TA-4G
2nd Batch (deliv 08/71)		
N13-154647	878	TA-4G
N13-154648	879	TA-4G
N13-155051	870	A-4G
N13-155052	871	A-4G
N13-155055	872	A-4G
N13-155060	873	A-4G
N13-155061	874	A-4G
N13-155062	875	A-4G
N13-155063	876	A-4G
N13-155064	877	A-4G

In Australian service, the Skyhawks retained US Navy serial numbers preceded by the RAN serial prefix for the Skyhawk, N13. These were supplemented by a 'buzz' or side number painted on the noses of the aircraft.

All 20 Skyhawks were delivered in standard US Navy light gull grey and white colours. Decorative paintwork provided variations on this but a major change came in 1979 when most of the surviving aircraft were repainted in an overall pattern of aircraft grey and light admiralty grey camouflage.

RAN Skyhawks of 805 Squadron on board HMAS Melbourne *in October 1979 during Exercise Kangaroo Three in the Coral Sea. A Wessex helicopter is about to touch down in the background. (via Ross Gillett)*

A-4G Skyhawks '884' (top) and '877' (bottom) display the early and late, respectively, colour schemes used by the RAN on these aircraft. Both are on HMAS *Melbourne's* flight deck, one about to launch and the other just recovered. (via Ross Gillett)

HMAS Melbourne with her Douglas A-4G fighter/bomber air group aboard. Tucked away neatly in the shadow of the island is a lone Wessex helicopter. (Navy PR)

The RAN's A-4 Skyhawks provided a viable and highly mobile tactical strike force in their day. (bottom) While still wearing its RAN colours, this A-4G has already been transferred to the RNZAF. (Eric Allen & Mitch Lovett)

Royal Australian Navy Skyhawks on the ground at Nowra NSW: TA-4G '878' (top) and A-4G '884' along with TA-4G '881'. Both of the latter were sold to the RNZAF in 1984. (Eric Allen)

The RAN achieved excellent service from its small inventory of McDonnell Douglas A-4G Skyhawks. Top photo depicts a dual seat TA-4G at NAS Nowra while the lower photo captures a single seat A-4G with long-range tanks, circa 1975. (Eric Allen)

A-4G Skyhawk '884' (top) displays underside detail as it overflies the photographer with everything down, while '871' (bottom) is photographed on board HMAS *Melbourne* in November 1979. (RAN/Richard Hourigan)

TA-4G Skyhawk '881' (top) went to the RNZAF in 1984, while the RNZAF's NZ6213 (bottom) was the former RAN N13-154905/884. It was photographed at Darwin in April 1988. (Bruce Chapman/Bill Lines)

SKYHAWK IN SERVICE

The story of the McDonnell Douglas Skyhawk's service career with Australia involves much more than just the aircraft's time with the Royal Australian Navy Fleet Air Arm. As is discussed later in this chapter, the sale of the 10 surviving RAN Skyhawks to the Royal New Zealand Air Force in 1984 and the subsequent return to Australia of some of them (along with other RNZAF Skyhawks) under contract to the Australian Defence Forces forms a significant part of the aircraft's history 'down under'.

In addition, the Skyhawk's service with the RAN was noteworthy in a less positive sense as its demise resulted from the decision by Australia's political masters to retire the aircraft carrier and RAN flagship HMAS *Melbourne* and – after an interesting chain of events – not replace her. This action marked the end of operational fixed wing flying by the Fleet Air Arm and the end of the era of Australian carrier operations – "the end of civilisation as we knew it", quipped one former FAA officer to the author.

The arrival in Australia of *Melbourne* and her cargo of brand new A-4 Skyhawks and S-2 Trackers in November 1967 began the final era of RAN carrier operations. There had been several times over previous years when such operations had been threatened with curtailment by the government, but for now everything was in order for the foreseeable future.

Upon its return to Australia, *Melbourne* was immediately withdrawn from service to undergo a major refit at the Garden Island Naval Dockyard in Sydney. Costing $8.75m, the refit was completed in November 1968 and the carrier recommissioned in February 1969.

The refit was estimated by the Navy to have extended the life of *Melbourne* by a decade and involved major alterations to the ship's electrical, radar and fuel storage systems. The flight deck, magazines and hangars were also modified, a modified crash barrier was installed as was a new air conditioning system and the communications systems modernised.

In addition, the ship's hull and machinery was completely refitted and overhauled and a new mast was fitted behind the bridge. The result was a light carrier capable of extracting maximum effectiveness out of its new aircraft.

While *Melbourne's* refit was being carried out, the Skyhawks were based at Naval Air Station Nowra (HMAS *Albatross*) on the south coast of New South Wales where training got underway. The first operational Skyhawk squadron, No 805, was commissioned at Nowra on 10 January 1968.

The Skyhawks got their first taste of real carrier landings in November 1968 when full landings and then launches were made from the visiting British carrier HMS *Hermes*.

RAN Skyhawk Squadrons

Two Fleet Air Arm squadrons operated Skyhawks during the type's career with the RAN, 805 Squadron (later VF-805) as the front line unit operating from *Melbourne* when it was at sea and from Nowra when it wasn't; and 724 Squadron (later VC-724), the Skyhawk Operational Flying Training School, based at Nowra.

Fast and low, A-4G Skyhawk N13-155064/877 displays its lines and the aircraft grey/light admiralty grey camouflage introduced in 1979.

A-4Gs 888 and 889 both failed to survive operations from HMAS Melbourne, *the former as a result of arrestor wire failure in 1979 and the latter due to catapult failure six years earlier. (via Sid Finn)*

805 Squadron was the most senior of the two Skyhawk units having been recommissioned as an RAN unit in August 1948 after having previously existed as a Royal Navy squadron. 805 Squadron was equipped with Sea Furies in 1948, the aircraft being handed over in Britain and delivered to Australia aboard HMAS *Sydney* in February 1949.

The Sea Furies saw an intensive few months of active service in the Korean war over the late 1951/early 1952 period aboard HMAS *Sydney* (along with 808 Squadron's Sea Furies and 816 Squadron's Fireflies), the squadron retaining the Hawker fighter until 1957.

805 Squadron was disbanded at Nowra in March 1958 but reformed five days later as an all weather fighter squadron with de Havilland Sea Venom jets replacing the piston engined Furies for service aboard HMAS *Melbourne*.

The squadron was again disbanded in June 1963 and its aircraft absorbed into 724 Squadron but in January 1968 it was recommissioned with Skyhawks, again for service aboard *Melbourne*. It was redesignated VF-805 in 1969 (adopting the American system of naval squadron numbering) and remained equipped with Skyhawks until the end of RAN carrier operations forced what is to date its final disbanding in July 1982. The squadron's Skyhawks were transferred to VC-724.

724 Squadron also had its origins in a Royal Navy squadron which was formed at Bankstown NSW in April 1945 equipped with Beech 18 Expeditor and Avro Anson aircraft. The squadron was disbanded in May 1946 but reappeared nine years later in June 1955 as an RAN operational flying school for which role it operated a mixed fleet of Sea Furies, Fireflies, Wirraways, Vampires, Gannets and Sea Venoms.

In 1957 the squadron was equipped only with Sea Venoms and Gannets for its operational flying training role but the following year saw it concentrating on jet conversion and all weather fighter training duties with Sea Venoms.

In 1961 724 Squadron became a multi type unit again when 725 Squadron was disbanded and its aircraft absorbed into 724. The fleet included Sea Venoms, Gannets, Vampires and Dakotas and its role was changed to anti submarine and all weather operational flying training plus communications, trials and other duties as required.

In 1963 the aircraft from Nos 805 (Sea Venoms) joined the squadron as did the miscellaneous fixed wing types serving with 723 Squadron. By 1966 the squadron was operating as an all jet pilot conversion unit with Vampires and Sea Venoms, joined by newly delivered Skyhawks in early 1968. The squadron was redesignated VC-724 in 1969.

The Vampires and Sea Venoms were replaced by Macchi MB.326s in 1970 and the squadron's role was now defined as weapon training, photo reconnaissance and target towing as well as Skyhawk conversion training.

In July 1982 VC-724 absorbed VC-805's Skyhawks with the cessation of carrier operations and a year later the number in service had been reduced to four, mainly for target towing duties. In addition, the Macchis were transferred to the RAAF as RAN fixed wing flying generally wound down. The last quartet of RAN Skyhawks was withdrawn in June 1984 and the squadron was disbanded two months later.

Pilots converting to the Skyhawk at 724 Squadron undertook a six month course during which they typically flew about 110 hours on both the two seat TA-4G and single seat A-4G. A pilot's first Skyhawk solo usually took about five hours dual instruction to achieve and after 25 hours he was fully converted to the new type.

The remaining 85 hours of the course comprised the advanced phase, during which the skills necessary for operational flying – day and night air-to-air refuelling, tactics, use of the aircraft's terrain avoidance radar, interceptions, bombing, strafing, air-to-air firing and so on – were taught.

Some 100 mirror assisted simulated deck landings were then practised at Nowra before the new pilot tried the real thing on *Melbourne*, about 20 of these (and catapult launchings) being performed in daylight before the more demanding night landings and takeoffs were attempted.

Tours of Duty

Melbourne's Carrier Air Group (CAG) at the time of the first embarkation of the Skyhawks in 1969 comprised 805 Squadron (Skyhawks), 816 Squadron (Trackers) and 817 Squadron (Wessex helicopters). The normal mix of aircraft comprised four Skyhawks, six Trackers and eight Wessexes, although this could be varied. The maximum number of Skyhawks carried at any one time was eight, all of them single seaters due to of the fact that the two seat TA-4Gs were not permitted to operate from the carrier for reasons mentioned in the previous chapter.

Previously, *Melbourne* had normally carried 22 aircraft (10 Sea Venoms, 10 Gannets and two helicopters) but the much greater operational effectiveness of the new types more than made up for the later, smaller numbers.

The first commanding officer of the CAG was Commander Bryan Matthews, who had under him a staff of about 75 officers and 250 sailors making up the headquarters staff, support and squadron personnel required to the group running effectively. *Melbourne's* overall crew complement was about 1,200 officers and men.

A summary of *Melbourne's* career between 1969 and her decommissioning in June 1982 gives some idea of the variety of flying available to Fleet Air Arm pilots and the differing conditions they faced on a day to day basis as the ship moved around the world.

1969 saw *Melbourne* take part in the SEATO exercise Sea Spirit, an event which unfortunately ended in tragedy when she collided with the American destroyer *Frank E Evans* in the early hours of the morning of 6 June in the South China Sea. Seventy-four of the *Frank E Evans'* crew were killed in the incident, for which *Melbourne* was blameless.

Needless to say, the popular press made the most of the situation, referring to the tragedy five years earlier when the carrier had collided with the Australian destroyer HMAS *Voyager* off Nowra with the loss of 82 lives from the smaller ship, and using the words "jinxed ship" often enough to make the basically ignorant (of such things) public and some politicians believe it.

After repairs, *Melbourne* continued participating in SEATO exercises later in 1969 and into 1970, and in May 1970 was involved in a rather less serious activity when she visited Japan to participate in Australia's involvement in the Expo '70 exposition in Osaka.

Another major refit was begun at Garden Island in December 1970, the carrier's aircraft being disembarked to their shore bases. The refit was completed in August

A 'bolter'. Skyhawk 885's lack of extended tailhook hopefully indicates the pilot intends performing a touch and go rather than an arrested landing!

With a Sidewinder under its starboard wing, 886 accelerates along Melbourne's *deck with the help of the catapult. With the stick held hard back, the aircraft goes from 0 to 110 knots in as many feet. (via Ross Gillett)*

Just off! An A-4G at the moment of departure from HMAS Melbourne *in April 1979. (via Ross Gillett)*

(left) Exercises with friendly nations form a major part of any country's defence readiness. This 1980 shot shows three US Navy TA-4J Skyhawks refuelling from an Australian A-4G. (right) One of the Skyhawk's assets was its 'buddy' refuelling capability. Here, two of 805 Sqadron's aircraft demonstrate the technique while the pilot of 885 looks on. (via Ross Gillett/RAN)

(left) Skyhawk 883 of 724 Squadron, the RAN's Skyhawk training unit. (right) 'Chloe', the heavy sled used to test HMAS Melbourne's catapult, gets a workout at Sydney's Garden Island Naval Dockyard in June 1976. Two of the RAN's Skyhawks were lost due to catapult failure. (via Ross Gillett)

A-4G N13-154910/889 of 805 Squadron over the south coast of New South Wales.

1971 and the aircraft re-embarked the following month. *Melbourne* then took part in a series of multi nation exercises off Hawaii.

Subsequent highlights of Melbourne's travels include to Subic Bay in the Philippines as part of an RAN Task Group for Exercise Sea Hawk and subsequent joint Australian/NZ/UK exercises; to Pearl Harbour in August 1972 for Exercise Rimpac 72 then to Japan, Milne Bay and the Philippines before returning to Australia in November for another refit at Garden Island.

This one was completed in July 1973 and in the following month the carrier was off to Hawaii again for Rimpac 73. Voyages to Fremantle and Singapore completed the year.

1974 began with a quick visit to California in March then in June it was back to more local waters for Exercise Kangaroo One in the Coral Sea, a major event involving air, sea and land forces from Australia, New Zealand, Great Britain and the USA.

The last few days of 1974 and the beginning of 1975 were taken up by the relief operation to Darwin following the devastation caused by Cyclone Tracy on Christmas Day. Then it was off to Hawaii again in March for Rimpac '75 before the carrier's aircraft disembarked once more for yet another refit, this one involving considerable modernisation and not being completed until June 1976. By this stage of her career, *Melbourne* had travelled more than 662,000 miles (1,065,000km) since commissioning in 1955.

The last few years of *Melbourne's* service life saw participation in Rimpacs 77, 78 and 80 and Exercises Tasmanex (New Zealand), Sea Eagle 1 and Kangaroo Three in 1979 plus Sandgroper 80 off Fremantle the following year. Another refit was carried out between July 1978 and February 1979.

By 1980 the Australian government had decided to replace *Melbourne*, although at the time no-one could foretell that when she was finally decommissioned in 1982, circumstances (including a change of government the next year) would see her not replaced at all and the end of Royal Australian Navy carrier operations would come about.

Operational Flying

The following gives some insight into a 'day at the office' for an RAN Skyhawk pilot. The information was provided by Cmdr Keith Johnson RAN for Navy PR's Ross Gillett, who originally presented it in his book *Wings Across The Sea*.

It took a front line tour of 18 months, normally including two cruises on the carrier to make the novice a fully fledged FAA pilot. During the first cruise for the year, flying operations started slowly as the inexperienced adapted to the ship's routine.

After a week out from Sydney, the ship reached an operational tempo and the following day Skyhawk jets made four sorties of tactical reconnaissance and two night sorties of air intercepts.

The target for the first launch was to be an island 300 miles north. At the pilot's briefing, an army liaison officer explained that the

A 1978 shot showing Skyhawk 887 being prepared for launching from HMAS Melbourne. *A Grumman S-2 Tracker is parked in front of the superstructure. (via Ross Gillett)*

Pilot's eye view of HMAS Melbourne *with a just about full complement of aircraft on deck – eight Skyhawks, six Trackers and four Wessexes. (via Ross Gillett)*

island was used during World War II as a storage and refuelling depot for trans-Pacific flights. The state of the aerodrome and roads and the possibility of anchoring vessels in the harbour was to be assessed for intelligence purposes.

The flight profile was a high-low-high mission, with the last 100 miles to the island flown at 200 feet. A 10 mile error in track meant that the island could be missed. Morning brought a moderate swell and six-eighths cloud cover at 2,000 feet. Frequent showers with occasional thunderstorms and cloud buildups to 34,000 feet were forecast.

Black Leader briefed his Number 2 on the conduct of the mission: battle formation at height, strike at low level, 30 seconds separation over the island, 300 knots, one run. Pair rejoin for land on. In an emergency ejection the remaining aircraft was to circle the downed pilot and to guide in the rescue helicopter.

"Man aircraft" was broadcast. Maps and checklists were gathered and survival gear donned. On deck all was in readiness – both aircraft were ready to launch within ten minutes.

Number 2 is catapulted off successfully. Black Leader is then tensioned on the catapult. At the wind-up signal, the engine accelerates rapidly to full power. The pilot gives the 'go' signal and the green flag is lowered. Black Leader braces himself for the acceleration forces, head back, arm locked to restrain the stick, throttle held firmly forward.

The catapult releases, the aircraft reaches 110 knots in as many feet and is thrown from the bow. The pilot feels a slight sink as the wings strive for lift. Slowly, then rapidly as the undercarriage and flaps are raised, the aircraft flies away.

At 6,000 feet the Skyhawks emerge into bright sunlight and for half an hour steer accurate headings at 20,000 feet. They descend by radar, levelling at 200 feet above the waves. The target is positively identified and the reconnaissance completed. The return trip is uneventful.

For the jets' recovery, the carrier steers into the wind. The aircraft turn downwind at 15 second intervals to allow for 30 seconds separation on the deck.

By 130 knots, 450 feet up, undercarriage, flaps and hook are lowered. The reflected orange maze of the glideslope beam is seen in the deck mirror. The ship is rolling and pitching and Black Leader forces himself to believe the glideslope information from the landing aid and to ignore the gyrations of the deck.

In close, the deck pitches up, red lights flash and "wave off" is transmitted by the landing signals officer. Full power is applied and the jet flies away 30 feet above the deck. The landing officer signals that deck movement was too severe for safe landing.

Number 2 is in the groove and the deck is ready. The aircraft settles into the glideslope and the hook scoops up three wire. Number 2 is down, he clears the landing area quickly as Leader calls the 'ball'.

The approach looks good. The 'ball' sinks a little in close; he applies a little power. Then, violent deceleration as the hook snags a wire. The pilot hangs forward in his straps as the jet stops in 170 feet ...

General Duties

When embarked on *Melbourne*, the Skyhawks' primary roles were air defence, maritime strike and ground attack in support of the Army. The exercises mentioned above and other activities saw the aircraft working with the armed forces of many nations including the USA, United Kingdom, Canada, Indonesia, New Zealand, Japan and the Philippines.

The Skyhawk proved to be an effective asset, particularly in the sea and ground attack roles its ability to refuel in the air from other Skyhawks was a substantial plus, as was the large weapons load it could carry. Although armed with Sidewinder missiles for fleet defence duties in RAN service, its performance as a fighter was limited, with relatively poor manoeuvrability and performance at altitude.

While 805 Squadron was the 'combat' element of Australia's Skyhawk force and

With HMAS Hobart *in the background, a Skyhawk approaches HMAS* Melbourne. *At this stage the aircraft would be flying at about 120 knots. (via Ross Gillett)*

805 Squadron's full complement of eight A-4G Skyhawks lined up on Melbourne's *flight deck in early 1979, just before the aircrafts' colour schemes were changed. (via Ross Gillett)*

(left) Flight deck activity aboard Melbourne *in October 1974 off Jervis Bay NSW as Skyhawk 874 prepares to launch while Trackers stand by. Note the tall markings on the Skyhawk, red and white checks on a disc with superimposed black chess knight. This was one of the marking variations used by the 'Checkmates' Skyhawk display team. (right) The sign says it all, and just about everything the Skyhawk carried in RAN service is hanging off 885's weapons pylons! (via Ross Gillett/RAN)*

(left) Live armament firing at the Beecroft Peninsula weapons range. The Skyhawk could carry 5in Zuni or 2.75in FFAR rocket pods. (right) A-4G Skyhawk N13-154907/886 on board HMAS Melbourne *in 1979 with the 'planeguard' Wessex standing by. This aircraft was swept overboard later in the year. (Defence PR/via Ross Gillett)*

Flight deck scenes on board HMAS Melbourne*, the left shot being taken during exercise Kangaroo III in 1979. One of the tragedies resulting from the end of carrier flying in the RAN was the loss of skills which accompanied it. (via Ross Gillett)*

724 Squadron the training unit, the latter's aircraft also participated in exercises as well as providing operational works-up for RAN ships including that of target services. Most of this work took place off the New South Wales coast near Jervis Bay, although 724's aircraft were often deployed to RAAF bases around the country to support the ships of the Royal Australian Navy.

Accidents

By the end of the Skyhawk's 16 years of service with the RAN, exactly half of the 20 aircraft which had been delivered had been lost in accidents. Although a high proportion, the accident rate of aircraft operating from a carrier is always going to be greater than land based equivalents due to the very nature of the operations.

Interestingly, it is a spate of accidents in 1979-80 which contribute to the lion's share of the Skyhawk losses. During that period, six aircraft were lost (three in each year) including two in the same month, October 1980.

The RAN didn't lose its first Skyhawk until June 1973, some five-and-a-half years after the aircraft entered service. Another aircraft was lost in 1973 followed by single examples in 1974 and 1975 and then nothing for another 42 months until January 1979. No losses were recorded after 1980.

Below is a summary of the accidents resulting in the loss of 10 RAN Skyhawks:

N13-155060/873 (A-4G) 5 June 1973: Engine failure off Williamtown NSW; Sub Lt T der Kinderen ejected and picked up by RAAF Iroquois helicopter.

N13-154910/889 (A-4G) 8 November 1973: Catapult failure while launching from *Melbourne* during multi nation exercise near Singapore; Sub Lt B Evans remained in aircraft while ship went over it and freed himself underwater, picked up by helicopter.

N13-154648/879 (TA-4G) 16 May 1974: Crashed into sea 64km north-east of Nowra during mock attack on HMAS *Melbourne*, Lt R McMillan killed.

N13-155055/872 (A-4G) 17 July 1975: Crashed after collision with '870' near Beecroft Peninsula NSW firing range; Sub Lt K McCoy killed; '870' recovered, Lt Cdr G S King (CO of 805 Sqdn) unhurt; '870' suffered flame out, lost altitude and hit canopy of '872'; '870' landed into Nowra crash barrier after hydraulics failure and nosewheel collapse.

N13-155051/870 (A-4G) 23 January 1979: LT turbine failure 24km south-east of Braidwood NSW; Sub Lt C Tomlinson ejected and picked up by RAAF helicopter.

N13-154909/888 (A-4G) 23 May 1979: Crashed over side of *Melbourne* after arrestor wire broke during landing 90km east of Jervis Bay NSW; Lt Cdr K Finan (USN exchange pilot) rescued uninjured.

N13-154907/886 (A-4G) 24 September 1979: Rolled over side of *Melbourne* when ship suddenly rolled 20 degrees to starboard in heavy seas 355km east of Newcastle NSW; aircraft being repositioned at time, broke free of towing tractor and lashing chains; Aircraft Maintenance Sailor AB E Kann in cockpit suffered lacerations but managed to get clear and was thrown lifebuoy and joined in water by diver, both picked up by HMAS *Hobart*.

N13-154647/878 (TA-4G) 28 April 1980: Crashed 5km north of Nowra after engine failure while in the landing circuit at 1,000 feet; Lt A Sinclair ejected and picked up by RAN helicopter.

N13-155062/875 (A-4G) 2 October 1980: Engine failure during launch from *Melbourne* 70km north of Sumatra in Andaman Sea; Lt Cdr C Blennerhasset ejected safely.

Great shot of Skyhawk 885 catching the arrestor wire. This was the last of ten Skyhawks which were lost in RAN service, a catapult failure sending it to a watery grave in October 1980. (via Bruce Chapman)

N13-154906/885 (A-4G) 21 October 1980: Crashed off *Melbourne* during takeoff after catapult failure 200km south-west of Colombo; Sub Lt D Baddams ejected and picked up by Wessex helicopter. Final Skyhawk loss and second of that month.

The End of the Road

By the mid 1970s the Australian government realised that HMAS *Melbourne* would need to be replaced at some stage in the next decade. The search for a new carrier formally began in 1977 when companies were invited to register interest in the project, which called for a ship of about 20,000 tonnes loaded displacement capable of accommodating about 20 aircraft in a mix of anti submarine warfare (ASW) helicopters and STOVL (short takeoff/vertical landing) fixed wing combat aircraft, in other words British Aerospace Harriers.

Some 21 proposals were received from 16 overseas companies or consortia. Surprisingly, the British *Invincible* class (the nameship of which was launched in May 1977) was not on the short list presented in August 1979, those that were being primarily helicopter carriers.

Revised thinking led to the requirement being adjusted to include provision for Harriers (although the specific aircraft type was never mentioned – just the acronym STOVL) in September 1980, with the result that the *Invincible* class re-entered the equation a year later, a direct result of the British government's offer to sell one of these ships to Australia as it was intended that the Royal Navy keep only two of the three ships which were being built. HMS *Invincible* itself was commissioned into the Royal Navy in July 1980 with *Illustrious* and *Ark Royal* following.

Price and availability were the deciding factors. In February 1982, Defence Minister Jim Killen announced that HMS *Invincible* would be purchased from Britain at a cost of $478m including modifications, spare parts and test and training equipment. Delivery was scheduled for late 1983 and in the meantime *Melbourne* would be withdrawn from service.

The decision to purchase *Invincible* was a two edged sword for the RAN Fleet Air Arm. On the one hand it guaranteed the continuation of carrier flying of some kind in the service but on the other meant that the FAA's current fixed wing fleet – the Skyhawks and Trackers – would very shortly have no role.

No mention was made of acquiring any Harriers for the new carrier, the Minister in fact saying that no early decision was to be made on STOVL aircraft. For the moment, *Invincible* would carry only Sea King helicopters for ASW duties. In RAN service the ship would bear the name HMAS *Australia*.

The end came quickly for FAA fixed wing flying. In early April 1982 it was decided to pay off *Melbourne* as quickly as possible and to amalgamate the FAA's fixed wing squadrons.

By the middle of the year it was all over. *Melbourne* was paid off with due ceremony on 30 June and relegated to the strategic reserve while on 2 July VF-805 (Skyhawk) and VF-816 (Tracker) Squadrons were disbanded. The Trackers remained firmly on the ground at Nowra and were eventually offered for sale while the Skyhawks for the moment flew on with VC-724 Squadron at the same base. Discussion on *Melbourne's* ultimate fate continued and included one short lived plan to convert her into a tourist attraction!

A bit of international co-operation with A-4G 888 aboard HMS Hermes *in October 1968 while* Melbourne *was undergoing her major refit following delivery of the first batch of Skyhawks. Note the Sea Vixen in the background. (via Ross Gillett)*

On 30 June 1983, six of the RAN's remaining ten Skyhawks were withdrawn from service, while the other four were permitted to fly for another year on target towing, radar and weapon calibration and other duties in support of the fleet. For a short time there was discussion about retaining the Skyhawks for ground attack duties.

The last official Skyhawk flypast took place on 30 June 1984 when one aircraft towing a Delmar target flew a sortie which enabled the frigate HMAS *Sydney* to test its Phalanx Close In Weapon System.

On the same date, a family day was held at Nowra and in appropriately bleak and soggy conditions the final public Navy fixed wing flypast was performed. Flying one of the Skyhawks was Captain John Da Costa, Director of Naval Personnel Services and former CO of 724 Squadron along with the current CO, Lt Cdr John Hamilton in TA-4G '880'. Years earlier, Da Costa had been the first RAN pilot to fly a Skyhawk.

But at least there was still the prospect of HMAS *Australia*, or was there?

A war and a change of government finally put paid to the Royal Australian Navy's chances of ever operating an aircraft carrier again. When Argentina invaded the Falkland Islands in the South Atlantic on 2 April 1982, Australia lost its second last chance to acquire a new carrier. When the conservative Liberal-National Party coalition lost the March 1983 Federal election to the Labour Party, the last chance went out the window.

HMS *Invincible* and her Sea Harrier aircraft were a vital part of the British victory in the battles to reclaim the Falklands, so much so that when the conflict was all over in mid June 1982, the British government realised the asset it had in its 'Harrier Carriers' and withdrew the ship from sale.

The much older HMS *Hermes* (a conventional carrier fitted with a 'ski ramp' deck for Harrier operations and itself a major player in the Falklands) was offered instead, but after much discussion and evaluation of the options it was decided in February 1983 to temporarily defer the decision. A month later the election was held, the government changed and on the very same day it was sworn into office the new Defence Minister, Gordon Scholes, announced that no aircraft carrier would be purchased to replace *Melbourne*.

As for the old flagship, she was gradually stripped of equipment and sold for scrap. Her last voyage out of Sydney was in April 1985 when she was towed by tug to Shanghai to be cut up.

But she didn't give up without a fight – the tug's towing equipment failed in heavy seas off Stradbroke Island, Queensland, and repairs had to be carried out in Moreton Bay where she was also grounded for a time before continuing her last journey.

Action of a different kind in HMAS Melbourne*'s hangar during 1978 with a telecast of the Don Lane Show. Don Lane is on the left above the 'applause' sign, while comedian 'Ugly Dave' Grey entertains the crowd, a mix of the ship's company and associated families and friends. (via Ross Gillett)*

A-4G 885 (top) photographed over Sydney from another Skyhawk and 874 (bottom) on short finals into Nowra. (via Ross Gillett/Kevin J Hughes)

Skyhawk 887 photographed in July 1982 wearing the late colour scheme and 'Checkmates' emblem on the fin. (via Ross Gillett)

The Kiwi Connection

But the Australian Skyhawks would fly on. After negotiations the ten remaining Skyhawks and spares were sold to the Royal New Zealand Air Force for $28.2m. The aircraft were ferried from Nowra to RNZAF Ohakea in three batches in July 1984 where they would be modified and overhauled before entering service with Nos 2 and 75 Squadrons and joining the RNZAF's existing fleet of A-4Ks and TA-4Ks.

The aircraft were given RNZAF serial numbers, and the tie up with RAN serials is reflected in the table below:

RAN/RNZAF SKYHAWKS SERIAL NUMBERS CROSS REFERENCE

RAN	RNZAF
A-4G/A-4K	
N13-154903/882	NZ6211
N13-154904/883	NZ6212
N13-154905/884	NZ6213
N13-154908/887	NZ6214
N13-155052/871	NZ6215
N13-155061/874	NZ6216
N13-155063/876	NZ6217
N13-155064/877	NZ6218
TA-4G/TA-4K	
N13-154910/880	NZ6255
N13-154911/881	NZ6256

Whereas New Zealand's A-4Ks sport a medium green/dark green/dark grey camouflage colour scheme, the former Australian A-4Gs retained their RAN aircraft grey/light admiralty grey colours during early service with the RNZAF as well as the A-4G/TA-4G designations.

Later, when all the RNZAF's 21 Skyhawks were upgraded to 'Project Kahu' (Maori for 'hawk') standards, the former Australian aircraft were redesignated A-4Ks and TA-4Ks and repainted in the standard New Zealand scheme. They also gained one of the external physical characteristics of the 'K' models – the larger 'squared off' fin.

Early modifications applied to the A-4Gs – before the 'Kahu' programme – included the fitting of a braking parachute and VHF communications radios plus some minor instrument panel and switch position changes. One Skyhawk was found to have serious wing cracks and after arrival in New Zealand it was quickly fitted with a new set.

One former Australian Skyhawk was an early casualty in RNZAF service. In June 1985 NZ6118 (ex '877') flipped onto its back while landing in heavy rain at RAAF Townsville during a deployment to northern Australia.

Serious damage was sustained to the aircraft's tail, canopy, nose and other components and the aircraft was flown back to New Zealand in a C-130 Hercules. It remained in storage for nearly three years until April 1988 when the decision was made to repair it.

17,000 man hours and nearly two years later the Skyhawk was in the air again with a new tail section obtained from the US Navy, new wings and rebuilt aft fuselage, nose

Skyhawk 888 – with airbrakes extended – formates on the camera ship over the south coast of New South Wales. (via Sid Finn)

A 'farewell' flight by the RAN's Skyhawks and Macchis over Port Kembla NSW in June 1983. Four aircraft were actually permitted to fly on, performing target towing and support duties for another year. (via Ross Gillett)

A Royal New Zealand Air Force A-4K Skyhawk about to touch down. Note the square top of the fin and the dorsal avionics 'hump', the latter feature disappearing with the Kahu update programme.

section and nosewheel. The Skyhawk was then upgraded under the Project Kahu programme.

This programme, which was instigated in 1987 and completed three years later, basically involves an avionics upgrade in the Skyhawks, although a rewinging programme was first completed, using the spares and jigs which came with the large stock of ex RAN bits and pieces.

The avionics upgrade is based around new navigation and attack computer systems, the incorporation of HOTAS (Hand On Throttle And Stick) technology to effectively manipulate the new systems and a head up display (HUD). Data from the nav/attack systems is displayed on two cathode ray tube (CRT) screens on the substantially redesigned instrument panel and is also fed to the HUD, the primary source of flight information and weapon aiming control.

All modern and effective stuff, giving a new lease of life to the aircraft which also underwent a complete rewiring while the opportunity presented itself. RNZAF Skyhawks can carry a variety of modern weapons such as AIM-9L Sidewinder 'all aspect' air-to-air missiles, Maverick air-to-ground missiles, laser guided bombs and unguided rockets. The pair of internal 20mm cannon is retained.

Externally, A-4Ks converted to the new configuration lost their distinctive dorsal avionics 'humps', while the Australian aircraft of course didn't have them in the first place.

Return of the Skyhawks

In July 1990 it was announced that a five year agreement had been reached between the Australian and New Zealand Defence Forces which would see RNZAF Skyhawks from No 2 Squadron based at Nowra to provide air defence support for the Australian forces.

The preamble to the agreement summarises the deal and the benefits to both parties: "The government of Australia and the government of New Zealand having regard to the tradition of close and friendly relations between them, and taking into account of the close co-operation between them in defence matters, desire to enhance this co-operation to mutual advantage through the provision of certain air defence support flying for the Australian Defence Force by Skyhawk aircraft of the Royal New Zealand Air Force".

The arrangement involves up to 800 flying hours annually by the normal complement of four A-4K and two TA-4K Skyhawks (including of course, some former RAN aircraft which must feel completely at home at Nowra!) comprising air support operations for Australia and conversion of RNZAF pilots, No 2 Squadron's normal role.

New Zealand pays for the first 400 hours and Australia the remainder, and the benefits to both parties are substantial – Australia gets an operational need filled and New Zealand gets to exercise with a modern, well equipped Navy, carrying out realistic mock attacks on ships.

RNZAF Skyhawks began flying from Nowra in February 1991 and have subsequently been involved in several exercises apart from their ADF support duties, among them dissimilar combat training with RAAF Hornets at Williamtown. They've also provided the star turn at several Australian air shows over the years, all part of the great PR the Kiwis have generated as a result of their local activities.

And best of all, the countryside around Nowra still rings to the sound of Pratt & Whitney J52 turbojets pushing examples of 'Heinemann's Hot Rod' around the sky!

The former RAN N13-155069/877 was an early casualty in RNZAF service. The aircraft was repaired after this June 1985 accident at Townsville Qld.

The modern cockpit of the RNZAF's Kahu updated Skyhawks makes an interesting comparison with the standard A-4G cockpit illustrated in the following chapter.

The former RAN Skyhawks purchased by the RNZAF in 1984 retained their Australian colour schemes and the designation A-4G/TA-4G until they were upgraded later in the decade. The aircraft shown here are A-4G NZ6216 (ex N13-155061/874) and TA-4G NZ6255 (ex N13-154911/880). (George Erikson)

Four ex RAN Skyhawks in RNZAF service and modified to Kahu standards: (a) NZ6211/N13-154903, (b) NZ6217/N13-155063, (c) NZ6212/N13-154904 and (d) NZ6216/N13-155061. (Bill Lines)

FLYING THE SKYHAWK

Cdre Norman Lee, RAN (ret'd) became a naval aviator in 1947 and in his 33 year career (which ended in 1980 as the CO of NAS Nowra) he flew some 25 military types including Sea Venoms, Sea Furies, Scimitars, Sea Vixens, Gannets, Hunters and of course the Skyhawk. Here, Cdre Lee gives his impressions of Ed Heinemann' s compact and efficient design ...

My first close look at a Skyhawk was in the early sixties aboard a visiting USN carrier berthed at Garden Island dockyard. The ship was operating A-4Cs and a visit had been arranged for a team of us from NAS Nowra.

My initial impression of the aircraft focussed on the size of the cockpit; it was extremely cramped with barely room to move, most unlike American aircraft which are usually generously designed in this regard.

I was to learn some fifteen years later when I got to check out in the RAN A-4Gs, that Ed Heinemann, Douglas' chief designer, had set out to design the smallest practical attack aircraft and that the cockpit was actually designed to accommodate the 95 percentile man. This is a statistical body size and shape into which 95% of men will fit. It covers important dimensions such as height, girth, arm length and thigh length. In the case of the Skyhawk, the latter measurement was perhaps the most critical of all because if you exceeded it, you would leave your knee caps behind on ejection!

By the time I got to fly the aircraft I was in my late thirties and had thickened a little in the girth. All my other measurements were OK but it was a battle to do the autopilot checks as the control box was located on the right rear bulkhead of the cockpit and you had to reach round and look at it to run the checks, particularly in the early stages of your conversion.

With a delta shaped wing of only 27ft 6in (8.4m) wingspan, there was no need for the usual wing fold associated with naval aircraft, however it stood tall at 15ft (4.6m) and had an overall length of 44ft (13.4m). The controls were powered, with an unusual structural arrangement for the rudder which had external ribs. This was actually a later modification to overcome rudder buzz.

The empty weight of the A-4G was just under 10,500lbs (4,763kg) with a maximum takeoff weight of 24,500lbs (11,113kg) which could be any combination of weapons or external fuel, a remarkable capability. It was powered by a J52-P-8A turbojet which developed 9,300lbs of thrust. The "wet wing" had a capacity of 1,600lbs (728kg) of internal fuel; in addition three drop tanks each of 1,000lbs (454kg) capacity could be carried giving the aircraft a ferry range of over 2,000nms (3,705km). Drop tanks fuel was transferred to the main tank by air pressure controlled by a switch in the cockpit. Internal fuel was reduced to 700lbs (318kg) in the two seat TA-4G trainer to

TA-4G Skyhawk 881 lines up for takeoff at Nowra. Pilots converting to the Skyhawk typically flew about 110 hours on single and two seaters before becoming operational.

compensate for the shift in centre of gravity (CofG) with the extra cockpit.

Slower elevator response at low speed associated with the altered CofG prevented the trainer from being flown from HMAS *Melbourne* as it had been calculated that there would have been insufficient deck remaining to be able to lift the nose in the event of a "bolter".

The aircraft had a relatively primitive arrangement for overcoming pitch up when the speed brakes were operated. This consisted of large springs interlinking with speed brakes and the elevator control run. When the brakes were popped, the system grabbed a large handful of elevator and shoved the nose down, independently of the pilot!

Nosewheel steering was provided by pressing a button on top of the control column, the only snag was that it had to be kept pressed which was a bit of a pain. Surprisingly there was no parking brake which could prove to be tiring if the aircraft had to be held on the brakes for any length of time. Neither was there a battery or the capability of a self start.

A major feature of the aircraft was the three quarter span leading edge slats which automatically "popped" as speed was reduced in the landing circuit or if high 'g' was pulled. In conjunction with the angle of attack indicator, they made the aircraft relatively easy to fly on the approach. The latter instrument provided a read out of the optimum angle of attack to fly without the pilot having to make allowances for weight and angle of bank. It indicated by chevrons pointing up or down whether to increase or decrease attitude, and a circle when on optimum. Being well and truly on the back end of the drag curve on the approach, the landing was a no flare crunch onto the runway with the excellent undercarriage absorbing the shock.

The aircraft had only one bad characteristic, which proved fatal on one occasion, and that was that it needed some 15,000ft to fall through should it be stalled on its back. It literally fell flat until the nose finally dropped sufficiently to regain control.

By good fortune I returned to NAS Nowra as Commander (Air) shortly after we had taken delivery of the Skyhawks and Trackers so I was ideally placed to check out in our new aircraft. After a couple of dual rides in the TA-4G, my subsequent solo flying was done in the single seater. My initial impressions of a cramped cockpit were confirmed and I always felt as though I had been squeezed in with a shoe horn. The view ahead was good but severely restricted rearwards. I don't think I would have been too happy in real air-to-air combat!

Apart from the cockpit it was a pleasant little aircraft to fly having quite well coordinated controls. It accelerated quite rapidly on takeoff, particularly in the clean configuration and rotation speed was achieved in a matter of seconds. It was very manoeuvrable at low to medium altitudes but was fairly useless above 20,000ft. One had to remember however that it had been

TA-4G 880 makes a steep departure. (via Sid Finn)

The standard Skyhawk cockpit (in this case the front one of a TA-4G) should be compared with the A-4K Kahu update cockpit shown in the previous chapter. (Bruce Chapman)

designed as a light tactical strike aircraft and not as a fighter, hence it would have been foolish to think of it as being in the high altitude interceptor league.

Control forces with the hydraulics out were fairly heavy and only gentle manoeuvres were really possible; I doubt that a successful weapons mission would have been possible under these conditions. We used to practise hydraulic-out low precautionary approaches (LPAs) which providing you got the aircraft settled down and properly lined up, didn't make for too much sweat.

What was exciting however was the high precautionary approach (HPA) which was an approach designed to effect a landing from a flameout. It was not dissimilar to the approach made by the Space Shuttle, 250kts and a very steep descent requiring a flare to be initiated at some hundreds of feet on short finals from which if you had got it right, a landing was possible.

It was pleasant to fly on instruments, the avionics fit included Tacan, an ADF and a very good autopilot which had stick steer. This latter facility enabled you to take over from the autopilot by just grasping the control column. When you let go, it would take over and hold the attitude in which you had left the aircraft.

Having flown Hunters and Scimitars I was not particularly impressed by the Skyhawk's performance but soon came to realise that it was a first class weapons platform and could carry a load far exceeding that of either of those two aircraft. This was confirmed in my relatively short time in the Skyhawk in which I managed to get in a reasonable amount of weaponry including bombing, rocketing and strafing with its two 20mm cannon.

It was on one such mission when doing a period of refresher from Navy Office where I was now Director of Naval Aviation Policy (DNAP), that I gave myself a good scare. I had an aircraft configured with two drop tanks and carried out the usual fuel transfer checks after start up. This consisted of switching on the drop tank pressurisation and then switching it off whilst a member of the ground crew took the tank caps off to check that the tanks had pressurised. For some reason I failed to remake the switch in my haste to set off for the range.

Halfway through the sortie, which was very much head out of the cockpit, I got round to doing an instrument check; to my surprise the main fuel tank reading was way down. A very hasty check of the cockpit soon showed why, but to my relief shortly after making the pressurisation switch, the main tank reading began to increase.

It's funny how the mind works, I wasn't so much worried about my skin, I was more concerned about the ignominy of the senior FAA staff officer being seen to have screwed it up should I have flamed out!

On the subject of weaponry, the Skyhawk was fitted with a fixed reflector gunsight, not dissimilar to that installed in fighter aircraft before the Second World War! This is all that had been needed in its true role as a strike aircraft, however we were using it as a poor man's fighter and needed a gyro gunsight for air-to-air combat with the two cannon.

As DNAP, I pressed hard to have a lightweight gyro gunsight installed and had almost succeeded when another round of defence cuts saw the demise of the project. It left me pondering the logic of having the whole of the Navy supporting the young fighter pilot as he engaged the enemy, only to have him miss because of the lack of an adequate gunsight!

Again while in Navy Office I developed a somewhat cynical theory that projects which originated at the grass roots level decelerated in their passage upward through the system while those that were dreamt up at the highest levels accelerated downwards towards implementation. Never was this truer than in the case of the acquisition of the second ten Skyhawks. For some reason, which I was not a party to

RAN Skyhawks at Nowra circa 1979 with both original and new paint schemes evident.

as I was not in the chair at the time, Defence found itself in the embarrassing situation of having underspent for a particular year. The word went out from the Minister that anyone having a project ready to run would stand a good chance. It so happened that Navy had a plan set to go to acquire an additional eight second-hand single seaters and two more trainers from the USN.

Never has a project received a more rapid nod, in the normal course of events it would have taken years to fight it through the system but in this case it was done almost overnight. I trust that the procurement procedures are a little more sophisticated now!

Returning to the aircraft itself, in addition to weaponry I checked out at night and on my solo ride discovered a phenomenon not mentioned in any briefing. This occurred if the aircraft was being flown in very still conditions at a certain power setting, when a set of bleed valves associated with the compressor would momentarily open. The effect was very off putting as it felt like a transient loss of power. The type of which that makes you say in panic "what the hell was that?" When I mentioned it on landing the reply was "Sorry, forgot to mention that one!"

The only other aspect of the aircraft worth mentioning was its capability for air-to-air refuelling, something Navy had several decades before the RAAF. I managed to do both a tanker flight and a mission as receiver, neither of which presented any problems. The briefing for the latter was fairly simple: "Just fly your right foot into the drogue".

It was very sad when the Skyhawks went to the Kiwis. Most of us remember the throw away line, "You can forget about that", made by the late PM when asked whether his new government would proceed with the plan to acquire HMS *Invincible*. Seemed to make all those years building up a first class Fleet Air Arm worthwhile.

There is no doubt that the Skyhawk served the RAN well. It had the capacity to project medium level airpower, appropriate for our neck of the woods, and it could do it on the spot, twenty four hours a day. The fact that the Kiwis are still flying out 25 year old aircraft (albeit highly modified and extremely capable) shows what a solid and dependable machine it is. I count myself lucky to have been in the right posting at the right time to have been able to check out in it.

After about 100 simulated deck landings on dry land at Nowra, the new Skyhawk pilot performed about 20 real landings (and launches) as part of his conversion training. This pilot has just caught the wire.

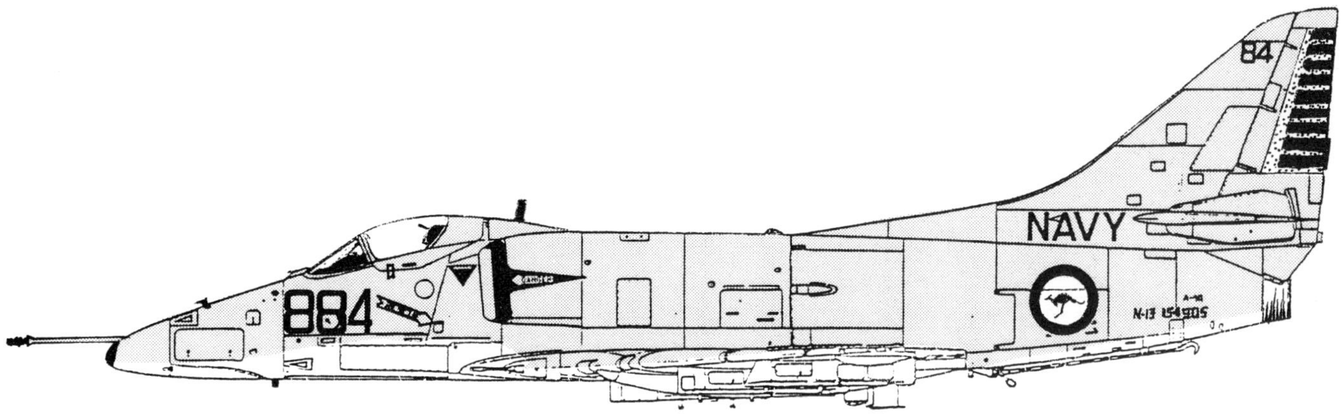

Skyhawk A-4G N-13 154905/884 of VC724, 1968
Standard scheme of gloss gull grey with gloss white undersurface. Insignia red edging on intake lip with white arrow and red lettering. Yellow rescue arrow below cockpit with black lettering and outline. Insignia red ejection warning triangle below cockpit. Black nose number, serial, NAVY, nose tip and antennas. Note the square style of the nose number on this particular aircraft. Insignia blue and yellow markings on rudder. Straight refuelling probe.

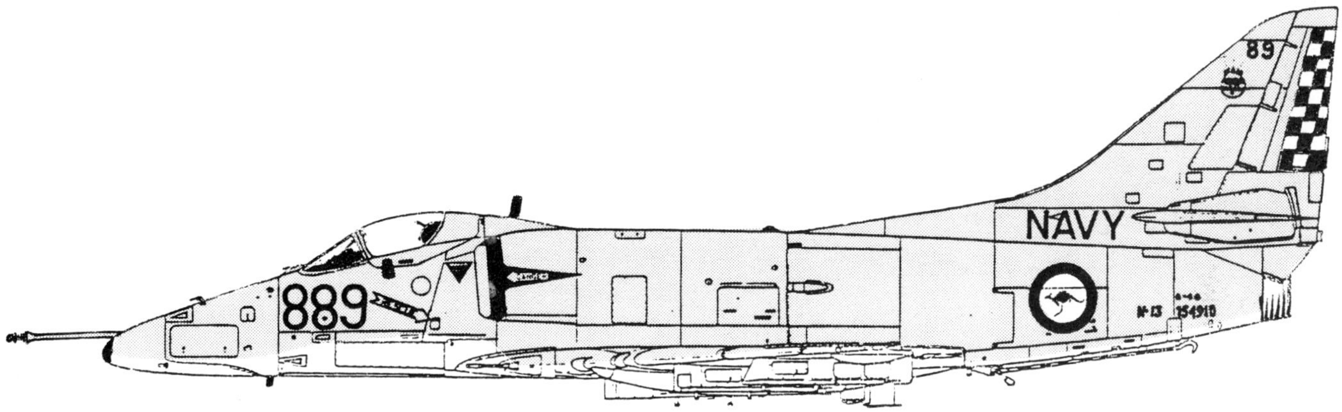

Skyhawk A-4G N-13 154910/889 of VF805, early 1970s
Standard scheme. Red and white checks on rudder. Squadron badge on fin.

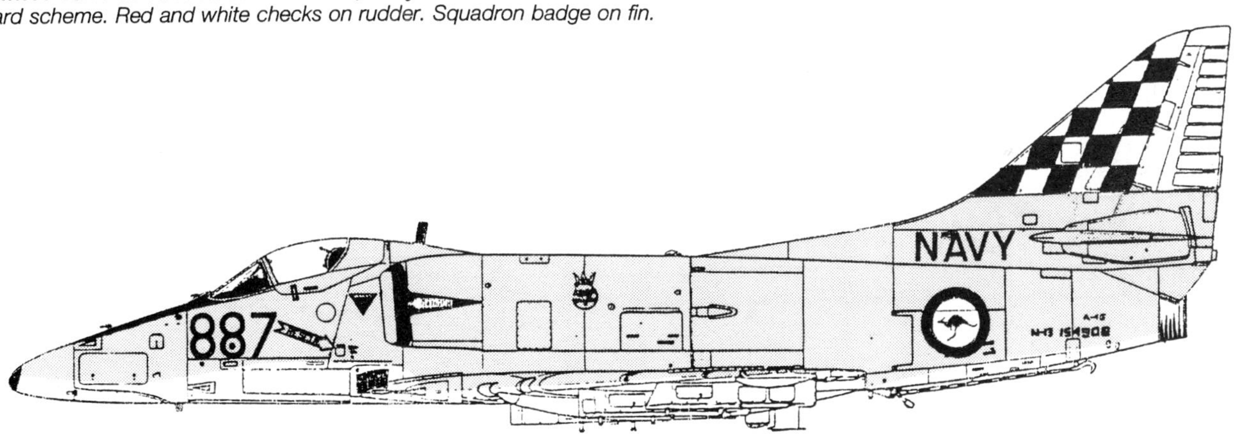

Skyhawk A-4G N-13 154908/887 of VF805, 1974
Standard scheme. Red and white checks on fin. Gloss white rudder. Squadron badge on fuselage. Black anti-glare panel forward of cockpit. Bent refuelling probe.

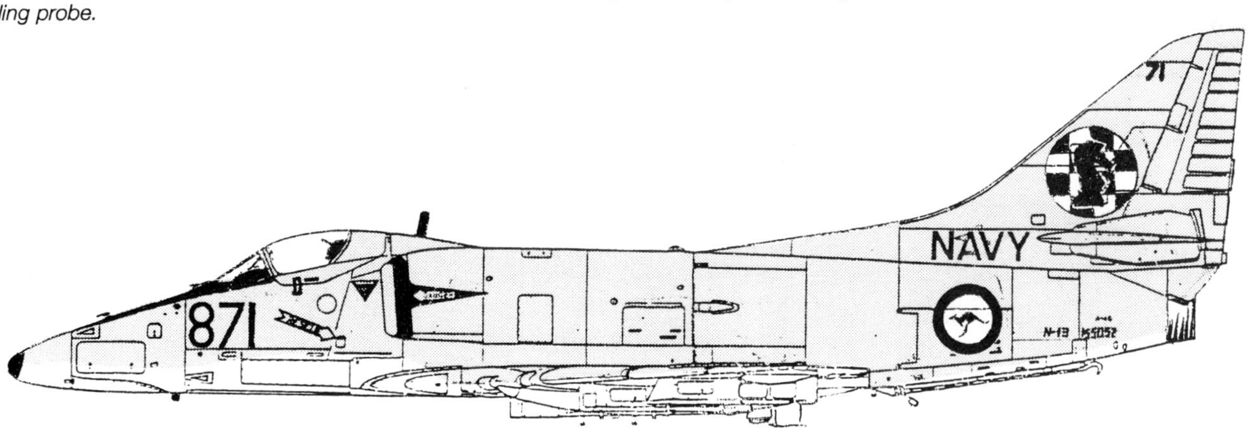

Skyhawk A-4G N-13 155052/871 of VC724, 1974, used by 'Checkmates' aerobatic team
Standard scheme. Tail markings comprise a black horse's head on a red and white checked disc. The head faces rearwards on both sides. Dark grey anti-glare panel. Cannon have been removed from wing roots.

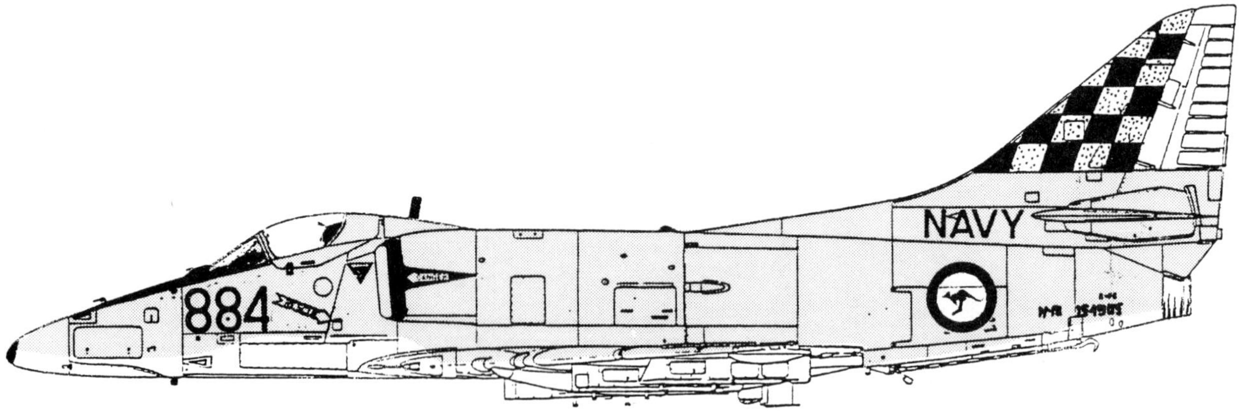

Skyhawk A-4G N-13 154905/884 of VC724, 1978
Standard scheme. Insignia blue and yellow checks on fin. Gloss white rudder.

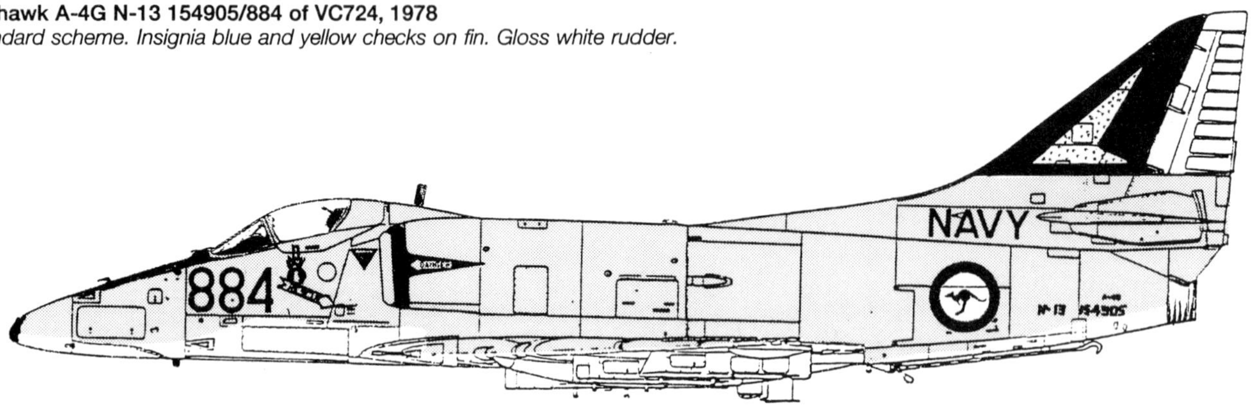

Skyhawk A-4G N-13 154905/884 of VC724, 1979
Standard scheme. Insignia blue fin with yellow flash. Gloss white rudder. Squadron badge below cockpit.

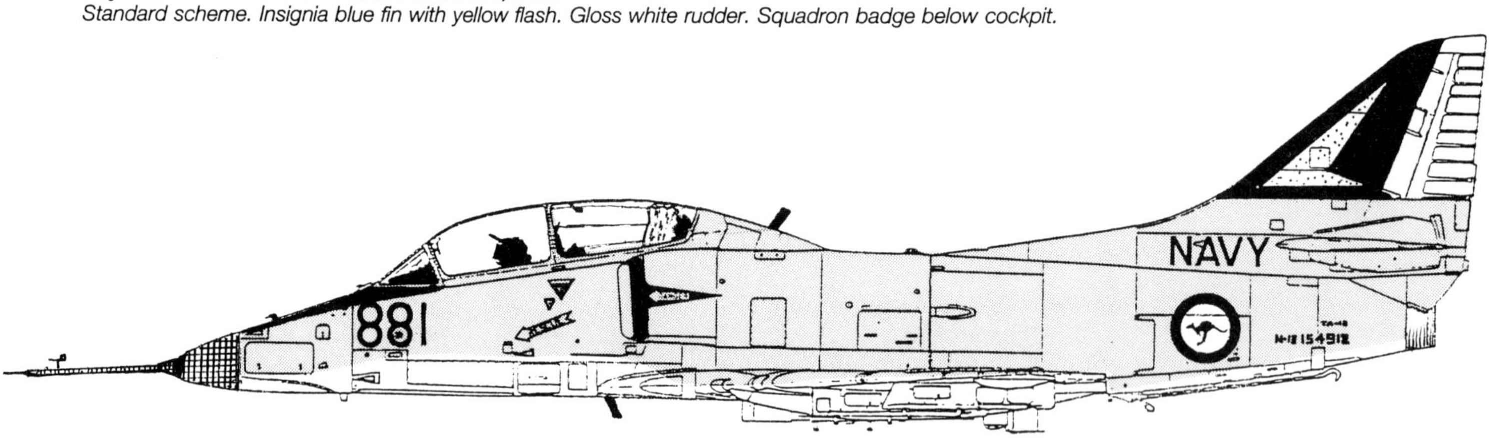

Skyhawk TA-4G N-13 154912/881 of VC724, late 1970s
Standard scheme. Insignia blue fin with yellow flash. Gloss white rudder. Orange nose. Orange nose probe with silver tip.

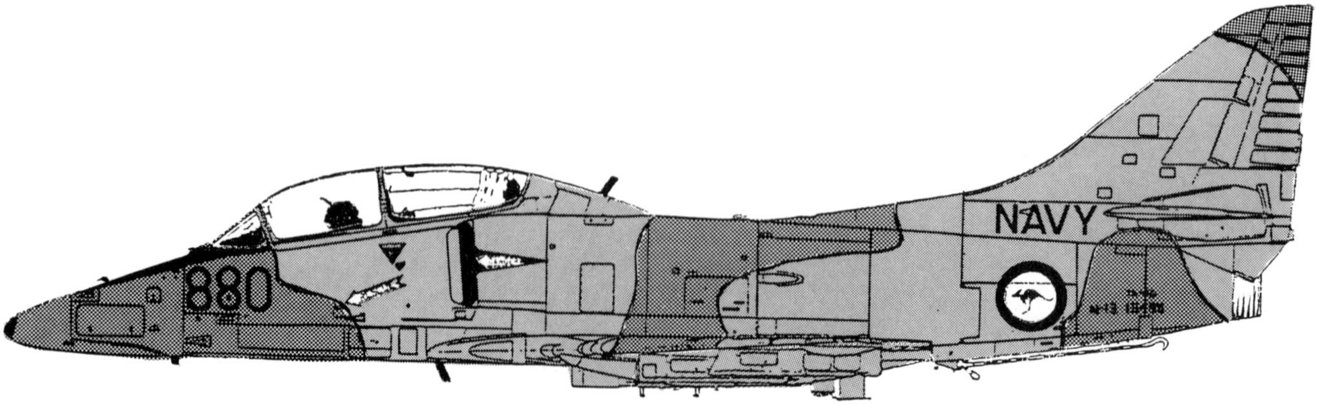

Skyhawk TA-4G N-13 154911/880, 1982
Wraparound scheme of Aircraft Grey and Light Admiralty Grey. All other markings are standard.

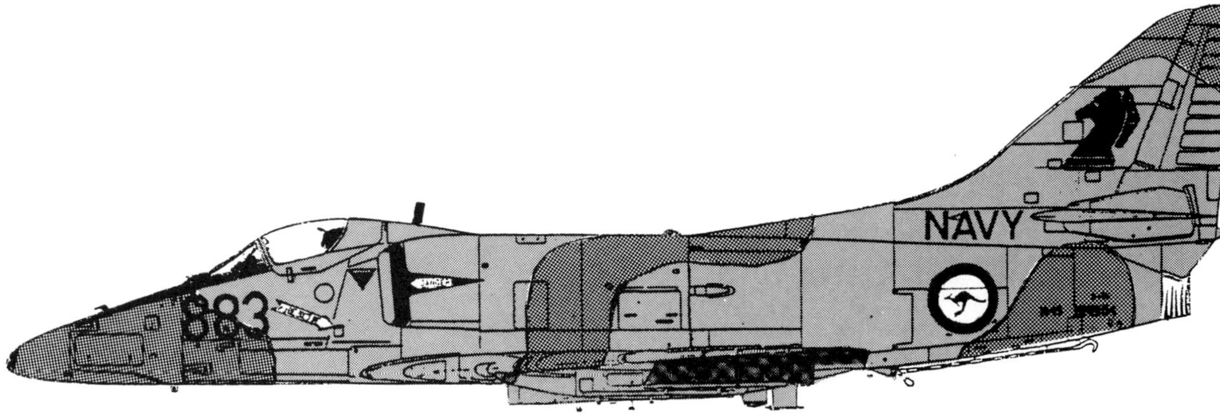

Skyhawk A-4G N-13 154904/883 of VF805, 1980
Standard wraparound scheme with black emblem on fin.

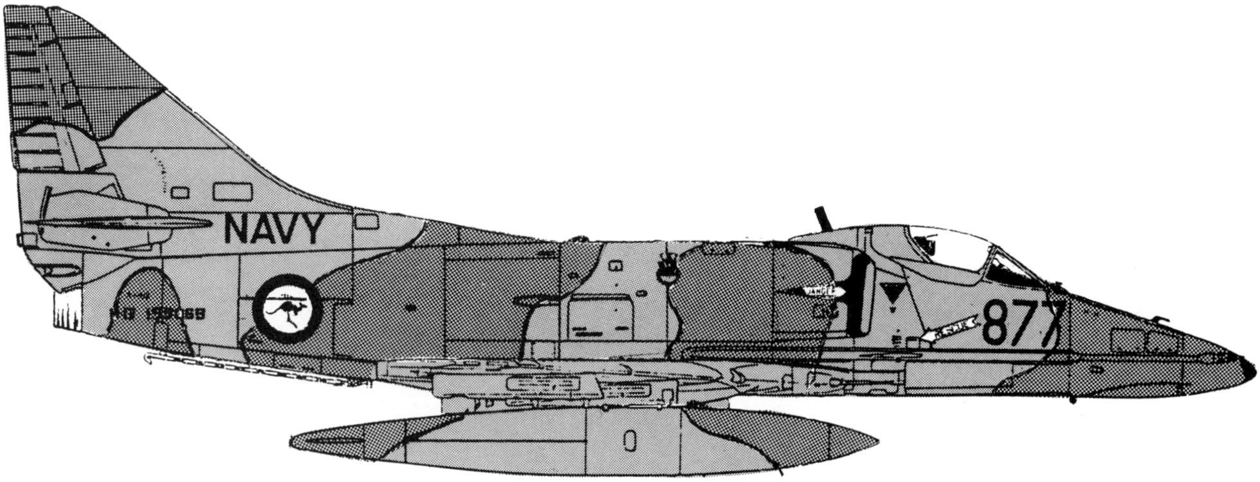

Skyhawk A-4G N-13 155064/877 of VF805, 1980
Standard wraparound scheme, showing the pattern for the right side. Squadron badge on fuselage.

Detail of tail marking of A-4G N-13 155052/871 (Illustration No 4).

Gloss Gull Grey (FS 16440)	*Yellow*	*Aircraft Grey (BS 381C-693 Matt)*	*Light Admiralty Grey (BS 381C-697 Matt)*

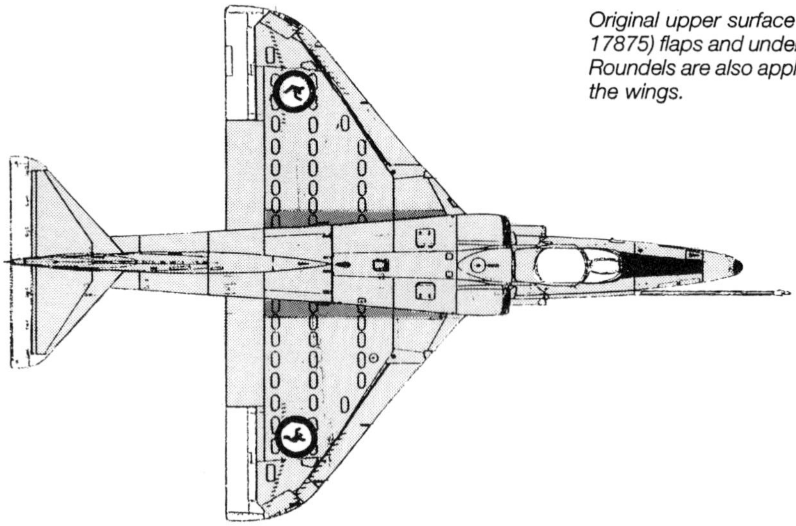

Original upper surface scheme of gloss gull grey with gloss white (FS 17875) flaps and undersurface. Note the dark grey walkways on wings. Roundels are also applied at the same location on the undersurfaces of the wings.

Upper surface pattern of the wraparound scheme.

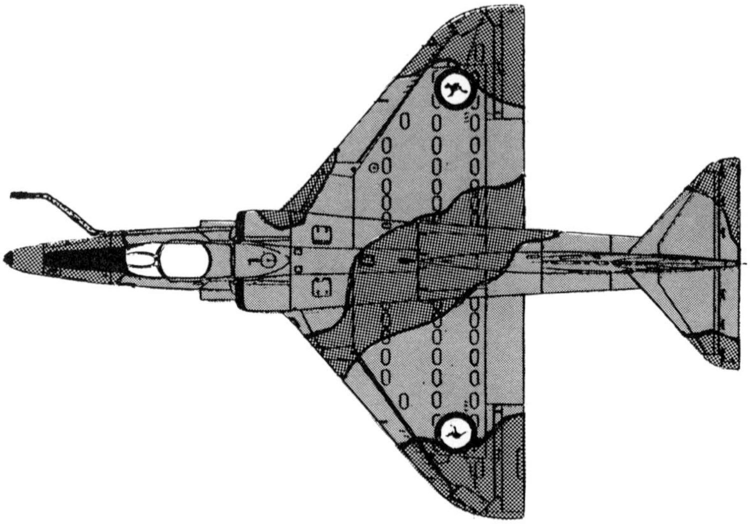

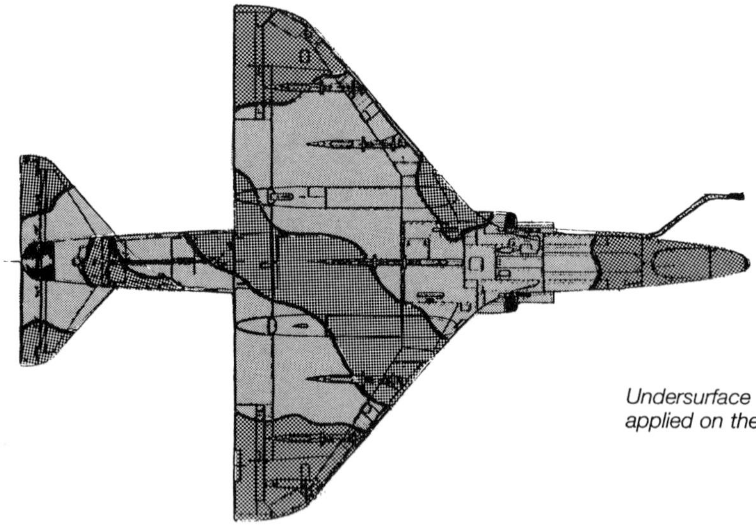

Undersurface pattern of the wraparound scheme. Roundels are not applied on the undersurfaces.

RAN McDONNELL DOUGLAS A-4 SKYHAWKS

Abbreviations: wfu – withdrawn from use; RNZAF – Royal New Zealand Air Force.

Notes: The RAN serial number reflects the Australian prefix (N13) and the US Navy number (the remainder of the serial). The 'Code' column refers to the 'buzz' number painted on the side of the aircraft. The 'Deliv' (delivery) column refers to the date the aircraft arrived in Australia.

RAN Serial	Model	Code	Deliv	Disposal/Remarks
N13-154647	TA-4G	878	08/71	crashed near Nowra NSW 04/80, engine failure
N13-154648	TA-4G	879	08/71	crashed into sea off Nowra NSW 05/74
N13-154903	A-4G	882	11/67	wfu 06/83, to RNZAF 07/84 as NZ6211
N13-154904	A-4G	883	11/67	wfu 06/83, to RNZAF 07/84 as NZ6212
N13-154905	A-4G	884	11/67	wfu 06/83, to RNZAF 07/84 as NZ6213
N13-154906	A-4G	885	11/67	crashed off HMAS Melbourne 200km S-W Colombo, catapult failure 10/80
N13-154907	A-4G	886	11/67	rolled off deck of HMAS Melbourne in storm 355km off east Australian coast 09/79
N13-154908	A-4G	887	11/67	wfu 06/83, to RNZAF 07/84 as NZ6214
N13-154909	A-4G	888	11/67	over side of HMAS Melbourne 90km east of Jervis Bay NSW 05/79, arrestor wire broke
N13-154910	A-4G	889	11/67	crashed off HMAS Melbourne off Singapore 11/73, catapult failure
N13-154911	TA-4G	880	11/67	wfu 06/83, to RNZAF 07/84 as NZ6255
N13-154912	TA-4G	881	11/67	wfu 06/83, to RNZAF 07/84 as NZ6256
N13-155051	A-4G	870	08/71	crashed near Braidwood NSW 01/79, LT turbine failure
N13-155052	A-4G	871	08/71	wfu 06/83, to RNZAF 07/84 as NZ6215
N13-155055	A-4G	872	08/71	crashed after collision with '870', Beecroft Peninsula NSW 07/75
N13-155060	A-4G	873	08/71	crashed 30km off Williamtown NSW 06/73 after engine failure
N13-155061	A-4G	874	08/71	wfu 06/83, to RNZAF 07/84 as NZ6216
N13-155062	A-4G	875	08/71	lost 70km north of Sumatra, engine failure during launch from HMAS Melbourne 10/80
N13-155063	A-4G	876	08/71	wfu 06/83, to RNZAF 07/84 as NZ6217
N13-155064	A-4G	877	08/71	wfu 06/83, to RNZAF 07/84 as NZ6218

IS AVIATION IMPORTANT TO YOU?

If so, then Australia's largest selling aviation journal can be your private consultant. Established in 1977, Australian Aviation is the largest selling aerospace journal in the southern hemisphere with a 96% paid circulation. That type of performance in a fiercely competitive and professional market does not come from being second best. If you need to know exactly what is happening in the fast paced world of aviation then Australian Aviation is YOUR answer.

One year (10 issues) mail subscriptions are available for $39.00 domestic. Overseas sea mail subs are $57 whilst overseas economy air mail rates are $60 for NZ and New Guinea and $80 elsewhere. All rates in Australian dollars. Or otherwise write requesting a free sample copy.

POST TO: AEROSPACE PUBLICATIONS PTY LTD
PO BOX 3105, WESTON CREEK ACT 2611
PHONE: (06) 288 1677, FAX (06) 288 2021

This is a: ❒ New Subscription ❒ Renewal

❒ I enclose my cheque/MO for $ .... payable to Aerospace Publications P/L or please charge to:

❒ Bankcard ❒ MasterCard ❒ American Express ❒ Visa Expiry Date ....

Card No: .... **Signature** ....

NAME ....

ADDRESS ....

.... **POSTCODE** .... **PHONE** ....

❒ ONE YEAR RATE: $39.00 (10 issues) – YOU SAVE $10.00

❒ TWO YEAR RATE: $69.00 (20 issues) – YOU SAVE $29.00

Overseas SEA MAIL rates are $57 (one year) and $105 (two years).
Overseas ECONOMY AIR MAIL rates are: NZ and New Guinea – $60 per year. Elsewhere – $80 per year.
All rates are expressed in Australian dollars.

YOUR SUBSCRIPTION WILL COMMENCE WITH THE NEXT AVAILABLE ISSUE. BACK COPIES ARE AVAILABLE AT $5 PER COPY POSTAGE PAID (O/SEAS RATES ON APPLICATION).

THANK YOU FOR YOUR PATRONAGE